The *Natural History* Reader in Evolution

THE *NATURAL HISTORY* READER
in EVOLUTION

NILES ELDREDGE, *Editor*

New York   COLUMBIA UNIVERSITY PRESS   *1987*

Library of Congress Cataloging-in-Publication Data

The Natural history reader in evolution.

Includes index.
1. Evolution. I. Eldredge, Niles.
QH366.2.N38 1987   575   87-5214
ISBN 0-231-06156-0
ISBN 0-231-06157-9 (pbk.)

Columbia University Press
New York   Guildford, Surrey
Copyright © 1987 Columbia University Press
All rights reserved

Printed in the United States of America

Clothbound editions of Columbia University Press books
are Smyth-sewn and printed on permanent and durable acid-free paper.

Contents

PART 6. EXTINCTION

PART 7. EVOLUTION AND SOCIAL ISSUES

Original Publication Dates of Articles
in Natural History Magazine

Evolution and Natural History

"Nothing in biology makes sense except in the light of evolution." So wrote the late geneticist Theodosius Dobzhansky, perhaps the central figure in the development of the "synthetic theory" of evolution, still the dominant set of ideas on how the evolutionary process actually works. In its broadest sense, biological evolution means that all organisms are descended from a common, primitive ancestor. All life is unified by virtue of its common origin. Life has achieved its present state of diversity as lineages of organisms have branched and become modified during the course of the last 3.5 billion years.

It was Dobzhansky who, in his book *Genetics and the Origin of Species* (1937), achieved the first comprehensive integration of the relatively young science of inheritance—genetics—with the original Darwinian vision of adaptation and natural selection. Critical to this venture was Dobzhansky's early training as a field naturalist, specializing in ladybird beetles. Darwin knew that organisms resemble their parents, and that there is heritable variation within populations. Both are necessary for natural selection to work, retaining favorable variations to maintain or improve the adaptations of organisms. In hammering out the essentials of an understanding of the principles of heredity, early geneticists began to question some of the older Darwinian notions. Thus mutations seemed often to be large in scale, and harmful in their effects—leading some geneticists to suppose that natural selection, working on a groundmass of smoothly intergradational variation, was not the real stuff of evolution. Large-scale mutations (if not lethal) were sufficient to lead to the instantaneous creation of new species—or so early saltational ("jumping") genetical theories of evolution saw the matter. But, by the thirties, geneticists had come to realize that many mutations are relatively slight in magnitude, and either neutral or downright beneficial in their effects—paving the way for a return to orthodox Darwinism, but this time with a firm basis in genetic principles. Dobzhansky went one step further, melding experimental results in the laboratory with observations of organisms in the wild.

Thus we have a theory of evolution that sprang in large measure from naturalists' observations. And, fifty years ago, the bond between genetics and natural history was strengthened. It would be quite reasonable, under these circumstances, to expect that the pages of such a magazine as *Natural History* would be chock-a-block full of articles on evolution, where biological phenomena are analyzed in detail to reveal the machinery of the evolutionary process. This book of readings selected from *Natural History* amply rewards this expectation. But, just as importantly, careful perusal of eighty-five years of publication of this popular journal of the American Museum of Natural History shows evolution to have been far more muted a theme than Dobzhansky's credo might lead any of us to expect. For evolution rarely appears in the magazine's yearly index (at least in the days before Stephen Jay Gould began his popular column *Ever Since Darwin* in 1974). It is as if evolution is in the background, only occasionally bursting forth directly as an author's central concern. Why evolution should have such a veiled role tells us much about the relationship between the basic idea of evolution and our contemplation of the living, natural world—as embodied in that charmingly archaic phrase, "natural history."

The theory of evolution that emerges in outline form through the introductions and contents of each of the seven succeeding parts of this book embraces organismic adaptation through natural selection, the origin of new species, and the rise and fall of major groups of organisms. Many other subsidiary themes are included as well. Evolutionary biology is almost labyrinthine in its diversity these days. No one biologist, nor even any single publication, can hope to master all its areas. Yet the practice of evolutionary biology—at least the job of elaborating theories on how the evolutionary process works—has largely fallen to the hands of geneticists ever since the birth of the "Modern Synthesis." Dobzhansky's tradition of melding nature with the laboratory, and theoretical work with observations of the real world, has been maintained—but remains difficult to accomplish in practice. The fact is, evolution is by its very nature historical. It is easier to simulate nature in the lab, controlling the variables and creating situations that induce genetic stability and change in experimental populations, than it is to extract evolutionary lessons directly from nature itself.

In the field, things are far more complex than in the laboratory. Organisms form associations—populations and species—which are usually spread out over large areas. And there is this uncanny feeling that the evolutionary game is somehow already over by the time the observer sets up camp: as we shall see in part 1, adaptation is the evolutionary explanation for design in nature—the apparently good fit between organisms and their environmental settings. Most organisms seem pretty well suited to their surroundings when we come upon them. The "evolutionary experiment" seems already to have been performed. Natural historians—embracing a disparate lot of systematists (who classify organisms), ecologists, paleontologists, and behaviorists—are left with the role of reconstructing the historical sequence of events that led to the present state of affairs in the organic realm. It is their lot to affirm that evolution has happened, and to write the story of life's evolutionary past. In general, such biologists have not been expected to concern themselves directly with evolutionary processes; it is geneticists who have borne most of the responsibility for the analysis of the actual mechanisms of evolution.

But it is not just the fact that evolutionary theory has remained strongly directed towards processes of genetics that evolution appears so sporadically as the basic theme of articles in *Natural History*. Nature, on a day-to-day basis, is actually organized into functional systems. The majority of practicing biologists in fact are "functionalists": they describe and analyze the structure and dynamics of biological systems, be the systems cell membranes, immune systems, circulatory systems, or ecosystems. Natural history falls easily into the realm of functional biology and this is the major reason why the majority of biologically oriented articles in the pages of *Natural History* seem almost a-evolutionary: their focus is on the workings of a single system, such as mating behavior in spiny lobsters, or foraging in mixed species flocks of South American birds. *Every one of these articles has evolutionary implications.* Most are concerned with describing adaptive features of organisms, be those adaptations anatomical, physiological, or behaviorial. It really does make sense to describe biological nature in functional terms, and it is possible to do so without overt reference to evolution. But to a reader with a Darwinian bent, the evolutionary message is always there to be gleaned in even the most straightforward of functionalist articles.

But better yet, a good number of authors writing for *Natural History* have taken that extra step and brought an evolutionary perspective to their field studies. The intersection of field-oriented biology (a perspective on the organic world that includes the nature and history of the physical environment as well as the organisms, species, higher taxa, and communities) with evolutionary theory is itself a complex affair, involving a variety of approaches and subdisciplines of biology. It is this relationship that forms the subject of this book.

The book starts off with a consideration of the core Darwinian tradition: adaptation via natural selection. "Neo-Darwinism" is that central body of theory that sees a genetic basis for most organismic features; variation from organism to organism in those features is likewise genetically based, and the ultimate source of that variation is mutation—spontaneous genetic changes. Because more organisms are produced each generation than can survive on existing resources (though we shall encounter some reasons to doubt that nature is actually so tightly organized in the essay by John Wiens in part 3), those organisms best suited to coping with life's exigencies will tend to out-survive, and, most crucially, out-reproduce their less gifted fellow organisms in the population. They will tend to leave more of their genes to the next generation. From the vantage point of a field naturalist, the concern with describing adaptations is basically functionalist. The questions asked are "How does this structure, or behavior, work?" "What is its biological purpose?" "What steps did the lineage have to go through to reach the stage we can observe right now?" Functionalism is merged with historicism in this particular application of an evolutionary perspective to observations in natural history.

Reconstructing evolutionary pathways and elucidating the "adaptive significance" of organismic features is a risky business—and recently has occasioned an outburst of self-criticism within the evolutionary community. The reason: evolutionary biologists have all too often given in to the temptation to make up "Just So" stories about how the elephant got its trunk, the giraffe its long neck—without casting their explanation in suitably rigorous ways that can be tested scientifically. But in this desire to sharpen up the evolutionary act, there has been a tendency to go perhaps a bit too far: we cannot lose sight of the fact that there *is* design in nature, that organisms *are* adapted, and that natural selection is the only known mechanism occurring in nature that can shape and maintain those adaptations.

Even more to the point, it is the ecologists, anatomists, behaviorists, and paleontologists who have the data and observations of organisms in nature that are the *sine qua non* of our very understanding of what adaptations are and how they have come to be. That there is a genetic basis of adaptations is undeniable. But the *study* of adaptations is far more the province of "natural historians" than it is an area that can be tackled in the genetics laboratory.

It was, once again, Theodosius Dobzhansky who initiated the modern study of the process of *speciation,* where new reproductive communities—*species*—become isolated from parental species, and begin to lead separate, independent "lives" as discrete, descendant entities. Part 2 is appropriately dedicated to this evolutionary phenomenon, as important a theme as is adaptation itself. Here, the field biologist and paleontologist comes into his or her real element, for although speciation requires an order of magnitude of thousands of years in most cases, a sampling of nature at any one moment reveals a spectrum of situations. Most species are fully formed, quite discrete and rather stable entities. But some seem more recently established: hybridization with close relatives may still be occurring, and divergence may not yet be complete. This spectrum from well-differentiatied single species on up through complete reproductive isolation, where two separate reproductive communities have formed from a single ancestor, forms the basis of our understanding of the factors governing the speciation process.

The theory of geographic speciation, where new reproductive communities originate through enforced geographic isolation, seems to fit the fossil record quite nicely. Indeed, fossils allow us to sample the entire history of a species, which can entail periods of ten million years or even more, as in the case of some common marine invertebrates. Data spanning such time periods add to the growing realization that species themselves are *individuals*—actual historical entities with their own births, histories and (eventually) deaths. From time to time, species may give birth to descendant species, adding force to the analogy between individual organisms and species. Such a perspective relies utterly on the field observations of natural historians for its empirical, real-world support.

Part 3 tackles behavior and ecology—two fields of biological study with strong functionalist traditions, but with obvious evolutionary overtones. The strength of the three essays included in this part, however, is the explicit link made by each of the authors between the data and theoretical perspectives of their own disciplines, with the larger theme of evolutionary processes. In perhaps the most controversial contribution in the entire collection, John Wiens casts strong doubt on the importance of competition between populations of different species living at the same time in the same place—"sympatrically." In several of the part introductions, I point to articles that use (or avoid) the notion in the elaboration of their themes. This collection clearly shows the importance evolutionarily minded biologists attach to interspecific competition.

With biogeography (part 4), we return to one of Darwin's original themes, as the distribution of organisms over the surface of the globe was one of the crucial items of evidence marshalled by Darwin to establish the very idea of evolution in the minds of his contemporaries. Once again we encounter a field that can be pursued for the most part without any regard to the notion of evolution. We can simply chart the present-day distribution of organisms; with fossils, we can compare the present with the past, and see how things have changed. But it is a profoundly richer exercise if we think of the effects that distributions —and their changes— have on the evolutionary histories of groups. Some authors writing in part 5 see adaptive radiations (proliferations of species, often with the invention of novel anatomical and behavioral features—new adaptations) coming as a reaction to modifications in geographic ranges. Colonization of new areas often leads to innovation. In other contributions, the strong relation between the physical environment, and in particular the shifting configuration of the major plates of the earth's crust, and the distributions and consequent evolutionary histories of entire major groups of organisms provide the central focus. With this topic, we begin to leave the realm of the properties of organisms and even of entire species, and deal instead with larger-scale natural groups, or *monophyletic taxa*. We become interested in the origins, histories, and ultimate fates of such large-scale evolutionary entities as families, orders, classes, and phyla.

Continuing on this level, parts 6 and 7 look at "living fossils" and extinction, respectively. Living fossils (organisms that bear a striking anatomical resemblance to the earliest members of their phylogenetic lineage, showing little change in some instances for hundreds of millions of years) actually constitute a problem in "macroevolution": large-scale evolution involves the origin of large-scale entities— the higher taxa such as Mammalia. Traditionally explained as long-term adaptive modification of a lineage, monophyletic taxa such as Mammalia are actually complexes of species. Patterns of speciation and coordinate adaptive modification of member organisms associated with the origins of such taxa are complex, and are best tackled in higher taxa with relatively few species. "Living fossils" typically come from long-lived lines with low rates of speciation and, concomitantly, low rates of adaptive transformation. Thus it is easier to assemble a relatively complete picture of the evolutionary history of the entire group, making such lineages attractive in the newly revivified corner of evolutionary biology that deals with the large-scale phenomena of "macroevolution."

Extinction, too, is of increasing interest to evolutionary theorists these days. Extinction is like colonization: by erasing much of the contents of the world's ecosystems, the evolutionary clock is reset, and life begins anew, with fresh bursts of evolutionary inventiveness based on whatever genetically based information manages to squeak through. We are abandoning the old view that evolutionary progress, via long-term adaptation through natural selection, is inevitable, built right into the evolutionary system. The currently more popular view sees stability of the world's ecosystems and conservatism in the histories of species and higher taxa. Only when the biological world is literally shaken up, when normal "background" extinction gives way to wholesale, across-the-board, cross-genealogical extinctions, does the opportunity present itself for the truly new to once again appear on the evolutionary scene.

The book ends with a look at evolutionary biology in a wider context. Evolution remains a troublesome idea to some segments of twentieth-century society—just as it looms as one of the very greatest ideas in the intellectual history of mankind to others of us. Included in the final essays are two excellent examinations of the modern version of creationism, the biblically based version of the origin of the universe, earth, and life, and the subsequent history of life on earth. With these essays, we complete the circle, as many of the moral and ethical ramifications that some people see in evolution were very much the subject of debate in Darwin's day.

I am convinced that the large-scale, "natural history" side of biology is more relevant to an understanding of evolution than it ever was before. With Darwin, who was forced to work without a clear understanding of heredity (actually, his own ideas on the subject were far off the mark!), "natural history" was about all there was available to serve as demonstration that life has evolved, and to provide the primary basis for understanding how the evolutionary process works. With genetics, and the more recent advances in understanding of the molecular biology of the gene, we have filled up some big holes in our biological knowledge; in so doing, we have tended to see evolutionary mechanisms predominantly in terms of genetic entities and phenomena. But biological "natural history" in the meantime has become the modern sciences of ecology, systematics, paleontology, and behavior. As Dobzhansky told us, it is evolution that informs all of biological knowledge, that enriches our understanding of biological nature at all levels of its complexity. The twenty-seven essays of this book show us just how the diverse subject matter of "natural history" really does bear on the understanding of the evolutionary process.

Niles Eldredge

The *Natural History* Reader in Evolution

PART 1

DARWIN'S HERITAGE: ADAPTATION AND NATURAL SELECTION

"Design in nature" constitutes one of the oldest observations in natural history. Surely even the inspired cave artists of Europe's Upper Paleolithic had a feeling for how well those bison, horses, deer, and ibex seem to fit their surroundings. In the mid-nineteenth century, Charles Darwin substituted a mechanistic theory of adaptation through natural selection for a biblically based story of divine creation as the explanation of organic design. Darwin saw selection-based adaptation as the inexorable motor of evolutionary change. It forms the central core of evolutionary theory to the present day—thus the rationale for devoting our first group of essays to this overarching evolutionary theme.

Sir Gavin de Beer was a distinguished, evolutionarily inclined British embryologist. His *Embryos and Ancestors,* first published in 1940, explored the connection between developmental and evolutionary phenomena. Here, de Beer adopts a clever device: were Darwin to revise his *Origin of Species* in 1966, what would he add? In answer, we receive a concise overview of both the history and basic content of evolutionary theory from this eminent biologist. And we find that to de Beer, as to so many of his contemporaries, evolution is virtually synonymous with the idea of adaptive change through natural selection. Indeed, he speaks of "the theory of evolution by natural selection of heritable variation."

That adaptation and natural selection form the very core of Darwin's evolutionary theory is made plain by the appearance of his name in each of the five essays of part 1. In the second essay, T. C. Emmel takes us to the field for a close look at that design in nature—in this case, in tropical forest butterflies. Along the way, he gives us a clear statement of the phenomenon of *mimicry,* where two (and sometimes more) species resemble each other very closely, yet are not particularly close relatives of one another. There are several types of mimicry, and Emmel gives us insights into the history of Batesian and Müllerian mimicry, revealing that it was tropical butterflies that supplied the original examples to Bates and Müller back in the nineteenth century.

With David L. Mulcahy's essay, we find yet another classic theme in evolutionary biology: the notion of adaptation through natural selection extrapolated to explain the origin and history of a major group (or *taxon*) of organisms—in this case, the flowering plants. This is the subject of macroevolution; it is a touchy area, and there have always been a few biologists since Darwin's day who have doubted that all large-scale evolutionary phenomena are simply the outcome of adaptive change extrapolated and extended over geological time. We shall encounter some of these doubters in later essays. Here we find a fascinating argument that competition and selection go on at many levels; Mulcahy is especially eloquent on the race of pollen to produce tubes for fertilization, which takes us into the realm of "sexual selection," the subject of the next essay by D. M. Waller.

Darwin spoke of "natural selection," contrasting it with "sexual selection." The distinction has become a bit blurred, as many biologists until comparatively recently saw no real difference between the two. But to Darwin, there was a world of difference between structures and behaviors shaped to facilitate an organism's survival (say, the echo-locating mechanism of bats) and features dedicated solely to reproduction, which commonly seem a detriment to an organism's survival (such as the elaborate plumage of male peacocks, or the aggressive behavior of males in many species, including our own).

In general, we should distinguish between *economic* adaptations (i.e., those concerned with making a living—mostly obtaining energy) and *reproductive* adaptations—attributes of organisms concerned solely with passing their genes along to the next generation, and, indeed, *creating* that next generation. *Natural selection* results when organisms within a population have differing amounts of success in the economic sphere of life—a differential that shows up, on average, in how successful they are in contributing genetic information to the next generation. *Sexual selection,* as Darwin first told us, is differential reproductive success based solely on reproductive adaptations. And that sort of relative success is responsible for the maintenance, and modification, of those very reproductive adaptations. Waller's article, detailing the complexities of the reproductive adaptations of jewel-weeds, takes us back to nature, and reveals the flexibility of jewel-weed's reproductive strategy.

Part 1 concludes with R. E. Cook's tale of adaptation in response to environmental disturbance. De Beer recounts the story of industrial melanism—the generally familiar case where selection has increased the frequency of dark moths, originally a rather rare morph, because soot from English factories darkened the bark of trees over the years. Industrial melanism remains the standard, textbook example of adaptive microevolution. Cook gives us an example from the plant realm—and a well-analyzed piece of research it is. Disturbed habitats, where the environmental challenge is great enough to provoke adaptive response, but not so great as to eradicate the local populations, are an intriguing blend of natural and experimental situations. Emmel, Mulcahy, and Waller describe nature; but in each instance they are discussing adaptations that appeared millions of years in the past. Their task is to reconstruct history. Cook tells us of a (necessarily more modest) case, but has the happy advantage to be sampling adaptive change as it is occurring. That it is rapid enough to be studied by humans is both fortunate and interesting: Darwin tended to preach the slowness, and gradualness, of most adaptive change. And Cook's article is valuable from yet another standpoint: Darwin based much of his understanding of the process of natural selection on what he learned from plant and animal breeders. At the end of his article, Cook performs the reverse operation: what we have learned from the botanical study of selection at work in an altered environment seems to have some direct implications for land management.

1

Darwin's *Origin* Today

GAVIN DE BEER

On January 16, 1869, Charles Darwin wrote to his friend Sir Joseph Dalton Hooker: "It is only about two years since the last edition of the *Origin,* and I am fairly disgusted to find how much I have to modify and how much I ought to add." On January 22, he continued, "If I lived twenty more years and was able to work, how I should have to modify the *Origin,* and how much the views on all points will have to be modified."

At that time Darwin was seriously troubled by two lines of attack on the *Origin* that appeared to be dangerous and damaging. One was a criticism brought forward by Fleeming Jenkin, who objected that the chances of single variations (that is, mutations) becoming incorporated in a population were infinitesimally small because of the infrequency (in the then current state of knowledge) with which two similar variants could be expected to meet. He also said it was virtually certain that such variants would be swamped and obliterated by interbreeding with the rest of the population.

Jenkin's criticism increased the difficulty under which Darwin was already laboring to account for a supply of variation sufficient for natural selection to work on. Darwin admitted to Alfred Russel Wallace on January 22, 1869: "F. Jenkin argued . . . against single variations ever being perpetuated, and has convinced me." In the new (5th) edition of the *Origin* then in preparation, Darwin did the best he could, which was to lean more heavily on the position that variation was produced as the result—then supposedly inherited—of acquired characters, of the use and disuse of different portions of the anatomy, and of environmental action.

The other attack that Darwin had to meet was from Sir William Thomson, afterward Lord Kelvin, who claimed that the rate of cooling of the earth proved that its age could not be estimated at more than forty million years. This was extremely damaging to the theory that evolution was caused by the natural selection of random variations, and was "opportunistic" because the time available would have been insufficient to allow for the evolution of all organisms from the primordial germ, unless design and direction had been at work. This was a basic and direct threat to Darwin's constant aim to keep the subject of evolution on a strictly scientific basis, free from metaphysical or theological concepts of providential guidance, which would, of course, have involved supernatural interference with the laws of nature.

That Darwin was shaken by this second blow is shown by a letter he wrote on January 31, 1869. "I am greatly troubled at the short duration of the world according to Sir W. Thomson for I require for my theoretical views a very long period before the Cambrian formation." But Darwin himself as a geologist had devoted prolonged attention to the length of time that must have been involved in the deposition of sedimentary rocks, and he felt justified in writing to Hooker on July 24, 1869; "I feel a conviction that the world will be found rather older than Thomson makes it."

From this standpoint Darwin has been triumphantly vindicated by the discoveries of radioactivity. Today the age of the habitable earth is estimated at some three thousand million years—ample for what Darwin called "wasteful blundering" and blind action of natural selection to have produced what it has. Darwin's fears on that score can be removed.

The Impact of Mendel

The manner in which Jenkin's attack has been parried may be introduced by quoting a passage in the *Origin* in which Darwin wrote: "The laws governing inheritance are for the most part unknown." Even the 6th edition, published in 1872, contains this passage, which, had Darwin (and everyone else) known it, was already overtaken by events. On February 8 and March 8, 1865, G. J. Mendel had delivered the famous lectures in which he laid down the foundations of the science of genetics, based on his work with generations of garden peas. Mendel's work remained unknown until 1900, when it was unearthed and confirmed, but even then the biologists of the day failed to appreciate its significance. Because the character differences then known to obey Mendel's laws were clear-cut, the opposition to Darwin's view of gradual and infinitesimal variation saw in Mendel's work a stick with which to beat Darwin.

It remained for Ronald Fisher in 1930 to show the real importance of Mendel's discovery, which was that inheritance is particulate—which means that variance is preserved instead of being "swamped," as had been assumed under the false notion of blending inheritance. Darwin, of course, never knew this, but he need not have worried on either of the scores that troubled him so greatly in 1869. The amount by which the *Origin* has had to be modified to keep it abreast of the present state of knowledge is much less than Darwin thought, from the point of view of theory, and there is much more evidence now available that confirms, extends, and refines its arguments.

Genetics and the Theory

Confirmation of the validity and reality of the principle of natural selection comes from two sources, genetic and paleontological. Taking the genetic evidence first, it was shown by Fisher that the phenotypic effects of a gene are subject to control by the other genes of the gene complex, and that as the gene complex is reshuffled in every generation by the segregation and recombination of the genes, the resultant individuals show variation of the effects of the gene in question. These effects can be *gradually* enhanced or diminished, according to which gene complex provides the most efficient adaptation of the organism to its environment.

This is why some advantageous genes have become dominant, and others have become recessive and even suppressed. E. B. Ford showed about twenty-five years ago that a given gene in currant moths can be made to become dominant or recessive according to the direction of the selection exerted on different lineages. In other words, there is incontrovertible evidence for selection at the heart of genetics. The phenotypic effects of a gene, clear-cut or not, are themselves the result of selection, and this selection gradually produces results—which is exactly what Darwin claimed.

Heredity is particulate, but this does not mean that evolution is discontinuous or "jerky." In other words, Mendelian genetics and the chromosome mechanism provide exactly what is required to explain evolution by natural selection. A new edition of the *Origin* would say, therefore, that the laws governing inheritance are now known, and that heritable variation arises from the random recombination of segregated, previously mutant genes.

Fisher then showed by a simple demonstration that all attempts to explain evolution as a result of inner urges, fulfillment of needs, effects of use and disuse, stimuli from the environment, orthogenetic trends, or other metaphysical concepts are doomed from the start. Such theories presuppose that there is a "favorable breeze of mutations" leading to adaptively directed and beneficial evolutionary results.

That such a process has no basis in fact is obvious. When a mutation first occurred, environmental conditions that then existed must have been adverse to the mutation. This is why the majority of mutant genes are recessive. This demonstration is so simple that it long evaded attention. But it is inescapable that, in the words of Fisher, "every theory of evolution which assumes, as do all the theories alternative to Natural Selection, that evolutionary changes can be explained by some hypothetical agency capable of controlling the mutations which occur, is involving a cause which demonstrably would not work even if it were known to exist." Genetics, therefore, shows that natural selection is all-powerful, while the immediate evolutionary consequences of mutation are negligible. It is only after mutant genes have been absorbed into the gene complex (if they become adaptively beneficial), segregated and recombined, and acted upon by selection that mutation plays a part.

The Zigzag of Evolution

From the paleontological side George Gaylord Simpson has shown that the rate of evolution is not correlated with variability, nor with the number of years occupied by a single generation. Furthermore, in the evolutionary history of such animals as horses, there has been no straight program at all. From the Eocene onward, the trends have zigzagged—first in the direction of many-toed browsers, then of many-toed grazers, and lastly of one-toed grazers. Those lineages that lingered and persisted too long in any of the previous trends of horse evolution paid the penalty of extinction. This, together with the demonstrable adaptation of successful lineages to changed ecological conditions, as revealed by geological and climatological data, shows that natural selection has been the governing factor in directing evolution. By comparing related marine and terrestrial animals it can likewise be shown that it is natural selection that determines whether evolution takes place rapidly, slowly, or remains stuck, because genetic mechanisms can produce either variability or stability—the former because genes can mutate, segregate, and cross over in their chromosomes; the latter because genes mutate only infrequently, never blend, and can be linked together in their chromosomes.

It is because natural selection is Darwin's personal contribution to science that his credit remains unblemished. It has sometimes been suggested that as he frequently spoke of "survival" as the prize of victory in selection, he was more interested in longevity than in reproductive capacity; and it has even been held that reproductive selection is "non-Darwinian." This is, however, unjustifiable, for by survival Darwin meant ability "to propagate their kind in larger numbers than the less well adapted." David Lack's demonstration that the *optimum* number of offspring for species survival is not equivalent to the *maximum* is relevant here.

Toward the end of his life, Darwin told his son Leonard that he expected evidence on natural selection to be available in about fifty years. As Fisher's and Simpson's work shows, this estimate was remarkably accurate, and the evidence now available is formidable and constantly increasing. However, only the most salient experimental results can be mentioned here.

Mimicry and Melanism

In 1936, E. B. Ford showed not only that Batesian mimicry (by which one species looks or acts like another) is a true adaptive phenomenon conveying survival value but also that it has been built up by natural selection of mutant genes. The proof is that where the models are more common than the mimics (in which case predators learn quickly to shun the unpalatable type), the mimetic resemblance is close to perfect, and the variance of the mimics is small. On the other hand, where the models are relatively infrequent, the mimics copy them only imperfectly and show considerable variance. When a model is less well known to predators, the survival value of resembling it is small, and there is less selection pressure exerted on the mimic to make it copy the model accurately. This is a case in which the close connection between genetics and ecology can be most easily observed.

A second example is furnished by industrial melanism, or color variations. In the middle of the nineteenth century in England, a melanic mutation of the peppered moth appeared, leading to the constant elimination of the melanic variety by bird predators because of its conspicuousness against the natural background of lichens on the trees where it rested. This was a telling case of adverse selection, but the mutation kept on recurring. With the progress of industrialization, the countryside became increasingly polluted by soot, and the trees became black. Now the original, gray, wild peppered moth suffered from bird predators in the industrial areas.

This phenomenon is widespread. More than seventy species of Lepidoptera are now undergoing melanization in industrial areas, and it has been observed in spiders as well. Here, then, is a gene that, when it first mutated, was deleterious but that, as a result of utterly unpredictable changes in the environmental conditions, became advantageous and now confers survival value. In fact, the degree of dominance of the gene has increased during the last hundred years. This is one case in which evolution has been under human observation, for the melanic form has been seen to supplant the old gray form in industrial areas, and natural selection has been shown to have directed the evolution.

Sickle Cells and Malaria

A final example is the mutation that causes sickle-shaped red blood cells among West African indigenes. The gene causes the formation of abnormal hemoglobin, the molecules of which attach themselves to one another end to end, thereby distorting the cells and causing them to look like sickles. These cells are easily destroyed, and in homozygotes (individuals that have inherited only the sickle-cell gene) under conditions of oxygen deficiency, this results in anemia, thrombosis, and death. It is not surprising that the gene is recessive. On the other hand, this abnormal hemoglobin prevents the entry into the red cells of the parasite *Plasmodium falciparum,* which is responsible for a type of malaria. In regions where malaria is endemic, an equilibrium is set up between the number of normal homozygous individuals liable to die of malaria and the number of individuals homozygous for the sickle gene that are liable to die of thrombosis. The heterozygous individuals (who have both normal and sickle genes) get the best of both worlds, for they are more protected from both dangers. But their genetic constitution makes inevitable the production of homozygote offspring of both kinds, who will pay their different kinds of penalties.

In West Africa, the sickle gene is present in about 20 percent of the population. With this percentage, four out of five homozygous sickle children die. The descendants of these populations in the United States, where there is no endemic malaria, show only 9 percent with sickle genes. This example shows how natural selection, opportunistically, can convert a lethal gene into one that confers survival value under certain ecological conditions. Furthermore, it provides a case of the special advantage enjoyed by heterozygotes, and shows how the percentage of a gene in a population can become changed. The latter is of particular importance because, as a result, evolution can also be defined as a statistical change in the gene pool of a population.

In this way, the theory of evolution by natural selection of heritable variation is established on an experimental basis to an extent that Darwin himself would hardly have imagined possible. Here, then, the *Origin* can be confirmed in theory, expanded in detail, explained in mechanism, and clarified. The same can be said of the fossil record, which by now has provided close series of lineages—in Jurassic ammonites, Cretaceous sea urchins, and Tertiary horses, camels, and elephants—and has also revealed forms that are indicators of the precursors of various classes of vertebrates and of the evolutionary stages intermediate between them.

Advances made in comparative anatomy and embryology since Darwin's day would fill in many chapters in a hypothetical new version of the *Origin*. For instance, I have found vestiges of egg-tooth papillae—similar to those of some reptiles—in embryos of marsupials, despite a hundred and twenty million years of viviparous reproduction. But references in the *Origin* to the Haeckelian theory of recapitulation (in which the succession of embryonic stages in a descendant directly represents the evolutionary stages of its adult ancestors) must be dropped in view of the much more satisfactory principle of pedomorphosis (in which lineages evolve from the youthful stages of their ancestors). Other advances fill out corresponding places in the *Origin:* F. C. R. Jourdain's study of mimicry in cuckoos' eggs; D. Lack's analysis of the taxonomy and ecology of the Galapagos finches (the birds that played such an important part in making an evolutionist of Darwin); and H. W. Lissmann's demonstration that weak electric discharges from muscles in fish can serve, on the principle of radar, to inform the animal of the proximity of other objects, thereby providing an explanation of the initial stages in the evolution of electric organs. The study of ethology at the hands of K. Lorenz and N. Tinbergen has revealed types of behavior that are adaptive and can be traced through related forms.

The *Origin* and New Sciences

To bring the *Origin* truly up to date, however, new chapters would have to be provided discussing branches of science that were not even dreamed of in Darwin's day. Here belong serology and immunology, which provide means of measuring the chemical divergence between the bloods and body fluids of different groups of related animals. Biochemistry shows that the affinities of animals can be revealed by the chemical substances built into their systems. Chromosome studies are another new field in which the minute investigation of translocations has enabled T. Dobzhansky to unravel the genealogy of some species of the fruit fly *Drosophila*.

T. H. Huxley, that rigorous puritan of science, always maintained that the final proof of the efficacy of natural selection as a cause of evolution and of the origin of species (not quite synonymous) would rest on whether it resulted in the production of reproductively isolated populations. K. F. Koopman has shown experimentally that *Drosophila pseudo-obscura* and *D. persimilis* are species that can interbreed, but that even so, matings between flies of the same species produce more offspring than matings between flies of different species.

Another topic that would have to be covered in a new *Origin* relates to population studies that may represent, as Ernst Mayr says, the most important recent revolution in biological concepts. In one sense, Darwin himself introduced population thinking, because instead of regarding a species as a "type," he stressed the variability of individuals within a species—"individual differences . . . frequently observed in the individuals of the same species inhabiting the same confined locality." But he slipped back into thinking of populations as types when discussing varieties and species. It is now necessary to realize that the product of evolution is a population with an adapted pattern of genetic inequality.

Sexual selection is a subject that received only brief mention in the *Origin* and, as Julian Huxley showed, it is in need of revision because some of the cases in which the sexes differ in structure, appearance, and behavior are not attributable to sexual selection, which benefits the reproductive capabilities of individuals of one sex, but to natural selection, which benefits the whole species.

The Gene Complex

Finally, to bring the *Origin* up to date, a new edition would contain a chapter of an agenda for the solution of chief outstanding problems, which are certainly no less numerous than when the first edition appeared. Such an agenda would necessarily include adequate theories of fitness, of sex-ratio control, of variation, and of how the effect of genes is under the control of other genes in the gene complex. This last problem will probably be worked out by the microbiologists—F. Jacob and J. Monod have already found that through chemically interrelated enzymes genes can collaborate to produce a controlling system that responds to changes in conditions. Most important, of course, would be the recognition that evolution must be considered as "dynamic" and not simply as "dynastic."

2

Adaptation on the Wing

THOMAS C. EMMEL

A lepidopterist's first day in a tropical rain forest is an unforgettable experience. My first such adventure came in Veracruz, a southern state in Mexico caressed by moisture-laden winds from the Gulf of Mexico. Walking among the green welter of giant, vine-festooned trees with their far-flung buttresses, I was more concerned with spotting snakes than butterflies. But soon all other thoughts were swept from my mind as I began to see the great number of gorgeous butterflies fluttering and gliding along the trail. The brilliant blue flashes of giant morphos, the tiny, iridescent green hairstreaks, the gaudily patterned undersides of the "88" and *Catagramma* butterflies, and the strange, loud clicking sounds broadcast by the *Hamadryas* nymphs as they flew from trunk to trunk combined to create an exciting moment for an impressionable student.

Since that day I have visited many rain forests throughout the American tropics, and every trip reinforces my belief that here is the true paradise on earth for students of butterflies. No other life zone has butterflies so diverse in structure and behavior, so numerous and visible, or so important a part of the community of plant and animal life.

Tropical rain forests straddle the equator in central Africa, the continental and island areas of Southeast Asia, the islands of the South Pacific, and northeastern Australia. But the most extensive rain forests of all spread a dense green carpet across Central and South America. These steaming jungles fascinated nineteenth-century European naturalists.

Such men as Charles Darwin, Henry Walter Bates, and Alfred Russel Wallace came to South America as young naturalists and left as seasoned biologists with a store of insights that would enrich the world. Darwin and Wallace in 1858 published their theories on the origin of species by natural selection and became famous. Less is known of Bates' contribution. In 1848, at age twenty-three, he set off with Wallace up the Amazon River to explore the natural history of Brazil. After traveling together for several years, they separated, but Bates stayed on in Brazil until 1859. He collected and sent back to England nearly 15,000 species of animals, mostly insects, of which more than 8,000 proved to be new to science. It is a record that will probably never be equaled. Bates' book, *The Naturalist on the River Amazon,* published in 1863, contained many of his observations on Neotropical butterflies. It proved an instant commercial success and stimulated a great interest in the natural history of the American tropics.

Aside from the large number of new species that Bates discovered, his most significant contribution was based on his observations of insect mimicry. While collecting on the lower reaches of the Amazon, Bates had noted uncanny resemblances in shapes, colors, and behavior between butterflies of very distinct families. He encountered many transparent-winged species of *Ithomia* butterflies floating in abundance in the shady ravines of the tropical forest. Now and then, flying among the *Ithomia* was a *Dismorphia,* also a clear-winged butterfly but one belonging to a totally different family, the Pieridae. Bates was unable to distinguish the two butterflies on the wing; each time he captured an *Ithomia* only to find it was a mimicking pierid, he could scarcely restrain an exclamation of surprise, so perfect was the mimicry in behavior, color pattern, and size. (It was later ascertained that ithomiids are protected by unpalatable secretions while the pierids are not.) This form of imitation has since come to be known as Batesian mimicry.

In 1859, after reading Darwin's *Origin of Species,* he saw that the most logical explanation was that an insect like *Dismorphia* improved its chances of survival by looking like a common unpalatable species. Bates concluded that the case offered a most beautiful proof of the theory of natural selection.

Fritz Müller came to the Brazilian Amazon a few years after Bates and discovered another important kind of mimicry among butterflies. While collecting ithomiids, Müller observed that a great many of these butterflies shared the same general color pattern, yet all of them were presumably unpalatable because their larvae fed on poisonous plants of the nightshade family. In his 1878 theory, which has become known as Müllerian mimicry, he suggested that sharing a common warning color pattern and common behavior was a survival advantage to each of the unpalatable species because predators learned to associate an unpleasant eating experience with that pattern; they had only to try one species of ithomiid to learn that eating it would produce such an unpleasant experience. Thereafter the predators would avoid all butterflies similar in appearance to the sampled one.

The tremendous variety of species in tropical regions has intrigued many biologists. Some indication of this variety is revealed in Bates' data. In all of South America (only partially collected even today), Bates became acquainted with 4,560 species of butterflies at a time when only 716 species were known to inhabit the entire Palaearctic region, from Europe to Manchuria, the best-studied area in the world. All of Europe contains only about 400 butterfly species, whereas within the radius of an hour's walk at Para (now Belem), Brazil, Bates easily collected 700 species.

The conditions that have produced such sharply contrasting figures are not yet completely understood. Biologists have observed the strikingly brief period of development for tropical insects. The danaid butterfly *Danaus chrysippus* usually goes through only one generation a year in the northern parts of its range in Asia, but in the southern Philippines it has a steady progression of generations throughout the year, each one taking only about three weeks to develop from egg to adult. The butterflies can accomplish this rapid turnover of generations because most tropical rain forests do not have well-defined seasons. Rainfall is a more or less constant quantity every month of the year, and the temperature range and length of day barely change. High temperatures and humidity, much light, and a great quantity of rapidly growing food make the tropics an ideal environment for cold-blooded butterflies. The additional generations possible over a certain period result in greater butterfly variation because more mutations can appear.

The biological environment of the tropical butterfly likewise contributes to creating diversity in these insects. The tremendous variety of plants available as hosts for the larval stages allows a much greater range of specialization on particular plants than is possible in the temperate zones, so there are many more opportunities, or niches, for tropical butterflies to exploit. Many plant species grow to certain heights in the forest and form discernible "layers," from the topmost canopy to the herb layer at ground level. The stratification of flight activity in different layers of vegetation increases the specialization opportunities available for additional species of butterflies.

The vast number of predators in the tropics also greatly influences butterfly variety. In their struggle for survival, many butterflies have adapted such devices as extensive mimicry of each other or, particularly in their larval and pupal stages, imitation of inedible objects, such as leaves or bird droppings.

A number of physiological and behavioral traits that serve as survival mechanisms have developed in the eighty or so morpho butterfly species of the American tropics. These traits are interesting examples of the evolutionary diversity, compared to arctic or temperate species, that is characteristic of butterflies in tropical rain forest habitats.

The upper surfaces of morphos' wings are usually brightly colored, ranging from white to light blue to dazzling deep blue to purple, although several species lack iridescence and flit about in somber yellow-spotted brown. The undersurfaces of all the species are brown, camouflaging them perfectly when they suddenly alight and fold their wings together. The slightly larger females are often less brilliant than their mates and have more retiring habits. Some, such as *Morpho cypris* and *M. theseus*, spend their lives in the treetops, while others, such as *M. peleides*, normally fly near the ground along paths and trails. Adults have been found to live close to nine months and apparently learn quite a bit about their environment. Many morphos follow regular flight routes during their daily movements through the rain forest.

In Central America, these flight paths apparently function as territories for male butterflies of *M. amathonte*. The male populations of this species form sleeping roosts consisting of a few individuals, which pass the night on top of large leaves within a few feet of each other. In the early morning hours, the males fly off to nearby feeding sites, which are accumulations of fermenting fruit on the forest floor. They usually feed between 7:30 and 9:30 A.M., without notably aggressive behavior. Females always feed after the males have left the fruits; thus no courtship occurs at the feeding sites. Between 9:30 A.M. and noon the butterflies move along somewhat circular flight paths. At this time, the males—each flying its own daily patrol route—attract, court, and mate with the females. This territorial behavior spaces out the males and draws them into a regular flight schedule through the forest understory, while the iridescence of their wings helps to attract females from afar.

Early in the afternoon, before the usual daily rains begin, the morphos, particularly the males, return to the general area of the feeding sites and roost there for the night. This daily activity pattern probably prevents aggressive behavior among the males when they are crowded together at feeding sites and nocturnal roosts. Their large but fragile wings would soon be battered by fighting if courtship and mating took place where several males were on hand.

One question may come to mind. If a morpho flaunting conspicuous wing coloration to attract females passes the same point on a trail at about the same time each day, why do not predators watch these flight paths and catch all the morphos that come along? A number of long-beaked birds such as jacamars, flycatchers, and motmots feed on butterflies, including morphos that are brown on both sides of their wings. But the brilliant blue species do not seem to be attacked very often. They seldom have beak marks on their wings where a bird tried to seize them, and their wings are rarely found in the piles of insect remains on the forest floor under the perches of these birds. Predators may learn to ignore these bright and showy butterflies since they are exceedingly difficult to catch. When chased by a bird, a flying morpho can quickly alter its pattern and speed of flight by increasing its characteristic vertical bobbing. As its mirrorlike iridescent wings flash in the sun, the morpho's flight becomes even more confusing to a predator, thereby almost inevitably insuring the butterfly's escape. A bird quickly learns that it is unprofitable to hunt this flashing, bobbing beauty.

Another diverse Neotropical rain forest group comprises the ithomiid butterflies—the 400 or so species characterized by a host of Batesian and Müllerian mimicry associations. The caterpillars feed on plants of the nightshade family, absorbing the poisonous alkaloids in the leaves. When passed along to the pupal and adult stages, these poisons continue to protect the butterfly against being eaten by a bird, lizard, or other predator. These butterflies, among other insects, may have evolved ways to detoxify the compounds they absorb, rendering them harmless to their own body tissues.

The life histories of the ithomiids have not been well studied, but those that have been traced reveal fascinating variations in the developmental stages between different species in this group. The larvae are often yellowish to green, with deeply creviced body segments; some resemble flattened species of common garden millipedes. Other larvae are cylindrical and encircled by black-and-white bands. In butterflies that lay clusters of eggs, the hatching larvae feed together as a group and move around communally for at least several instars, or stages between molts. Other species, which lay their eggs singly, share the solitary larval habits characteristic of most butterflies. The pupae of many ithomiids are a dazzling silver or gold, looking like large drops of forest dew or rainwater reflecting the morning sunlight.

All of the ithomiids exhibit a slow, fluttering flight, the slowest and weakest being that of the transparent-winged species. Because most of their wing scales have been reduced to hairs, leaving the membrane exposed, these ithomiids achieve a kind of transparency. Deep in the forest interior, moving over dark leaf litter in heavy shade, they are almost impossible to see or follow in flight. Many retain white marginal spots on their wings which may deflect a predator's attention from vital body parts. Other ithomiids take an alternate course and assume bright warning colors that advertise their unpalatability to predators.

Somewhat similar to the ithomiids in appearance are the long-winged heliconians, found only in the New World. These butterflies have specialized in eating the 400 or so types of passionflowers in the tropical jungles. Wherever these colorful plants blossom—in the forest canopy, in clearings, or along roadsides—

one or more of the eighty-odd species of heliconians are sure to be found. They are long-lived butterflies, some with life-spans up to six months, and like the ithomiids and morphos, they need to supplement their usual sugary nectar diet with nutritionally richer sources of nitrogen and vitamins.

The most advanced species in the genus *Heliconius* do not depend on accidental opportunities such as bird droppings but take advantage of a second nutritious component of the flowers they visit: the pollen. The *Heliconius* butterfly has a specially constructed tongue for collecting pollen. Nectar and enzymes regurgitated onto the pollen-laden tongue release free amino acids, which can then be sucked up and absorbed by the butterfly. These amino acids, the building blocks of proteins, are passed into the butterfly's tissues, and if it is a female, into her eggs.

Lawrence E. Gilbert, a lepidopterist at the University of Texas who has done considerable research with *Heliconius,* has found that adults of some species actually patrol a regular route each day between nectar and pollen sources over miles of forest, and that males of at least one species actively fight off other butterflies from the flowers selected for their daily early morning feeding.

Heliconians usually deposit their eggs on the very tips of the vine tendrils of passionflowers. Yellow-orange in color, these prominently placed eggs warn other females that this plant is already "taken." Thus a single vine does not become overloaded with caterpillars. At least one species of passionflower has orange tips on its vine tendrils, which could deceive a *Heliconius* female into thinking that the plant already has an egg on it. That particular passionflower vine may then escape being fed upon by *Heliconius* caterpillars.

Unlike the richly populated rain forests of South America, those of Africa contain fewer butterfly species. The African continent south of the Sahara (including the island of Madagascar) has just over 2600 described species (compared with over 6000 for South America), and a leading specialist in African butterflies, Robert Carcasson, estimates that this figure represents about 85 percent of the true total.

A majority of Africa's species are found in the rain forests and mountains of west and central Africa. Although many coppers and other lycaenids occur in the grasslands of south Africa, and other specialized groups reside in the rugged mountain ranges, it is the rain forest forms that are most distinctive in tropical Africa, as compared to the faunas of other world regions. The remarkable variety of trees, climbing vines, and other vegetation provides a rich source of food for caterpillars, and as in the American tropics, the small seasonal changes in the wettest rain forests allow for continuous reproduction.

The current drier climate, along with the influence of human cultivation and the extensive burning of the plains, has reduced the rain forests and separated the remaining stands by long stretches of grassland and sparse thorn scrub. As their habitat has broken up over the past several thousand years, such butterflies as the *Charaxes* species of the wet forests have formed different geographic races. Apparently the butterflies are unable to move from one forest area to another because of differences in humidity and other conditions in the intervening territory.

Among the tropical nymphalids, one attractive butterfly that is remarkable for its seasonal changes in color is *Precis* (a relative of the North American buckeye), whose many species are common throughout the woodlands and forests of Africa. Each species may have several color forms flying during the course of a year. In the dry season, for instance, the adults of *P. octavia* are heavily marked with black areas and blue spots, which may blend with the deep shadows of the sere vegetation. The wet-season form is smaller and bright orange with black markings. In *P. pelarga* populations, two dry-season forms appear. One is orange-banded like the wet-season form, while the other has an intense blue band. The wing margins of the dry-season forms in both *Precis* species are deeply scalloped and extended at the forewing tip and at the trailing edge of the hind wing. When the butterfly sits with folded wings among dead leaves, the false "stems" and the mottled browns and oranges on the underside make its resemblance to a leaf truly striking. Many of the satyrs also have seasonal forms, with more variable coloration and smaller eyespots appearing in the dry-season adults. All these changes in nymphalids and satyrs presumably aid in concealing the adult at a time when many plants in seasonal forest areas have lost their foliage and predators are avidly searching for food among reduced insect populations.

The largest butterfly on the African continent, the tailless swallowtail *Papilio antimachus,* is a rain forest inhabitant with an orange and reddish brown coloration bearing black markings. Males will occasionally visit mud at the edge of a forest stream, but their huge wingspan (more than nine inches) swiftly carries them away from danger if a predator appears. Many of the other Papilionidae have tails, which may aid in deflecting a bird's attack to this nonvital part of the hind wing. All are strong fliers, keeping to the forest or woodlands and often sailing high overhead among the canopy blossoms, with only an infrequent dip to a wet bank or understory flower.

One species has the common name of mocker swallowtail (*Papilio dardanus*) because many of the females mimic a wide variety of unpalatable butterflies. More than a dozen color forms are known, ranging from black and white, black and yellow, black and orange, and white and orange to nearly all black. The females of this mimetic species have even lost their tails, except in the races found in

Ethiopia and Madagascar, where they are tailed like the creamy yellow male. This extensive mimicry in the females probably became established because it allows larger numbers of the swallowtails to live in the same area. In Batesian mimicry assemblages, where edible butterflies imitate poisonous species, the mimic must usually be less abundant than the unpalatable model or else the predators will often try a mimic and associate its appearance with edibility. Mocker swallowtail females solve this dilemma by looking like several different species or subspecies of poisonous butterflies in a given region and thus parceling out the mimetic advantage over a number of color forms instead of one.

The remarkable family Acraeidae reaches its fullest diversity in the forest and savannas of central Africa. *Acraea encedon* in east and west Africa occurs in almost exclusively female communities in certain cultivated areas, an extraordinary situation that appears to be part of the way this species regulates its population. This is also one of the most varied butterfly species in Africa with as many as ten or more color forms. Four match the four African forms of the golden danaid *Danaus chrysippus,* a butterfly that has proved unpalatable to birds in feeding experiments.

Other *Acraea* species resemble each other, and many nymphalids seem to mimic acraeids. The complexities of these mimetic associations are great, and it is often not clear as to which species are the models and which are the mimics or just coincidentally similar in color pattern. Certain acraeine caterpillars feed on passionflowers, some of which are known to contain toxic compounds. Others, like *A. encedon,* apparently feed on nontoxic plants, or perhaps these butterflies do not derive the toxic compounds from plants but produce them metabolically.

The acraeids and their many mimics, along with *Papilio dardanus* and its diverse mimetic female forms, are providing key material to geneticists and students of ecology in unlocking the secrets of complex evolutionary histories. Combined with similar studies of New World butterflies, a deeper understanding of how plant and animal communities have coevolved is emerging. For a lepidopterist such as myself, that is reward enough, and the flash of a brilliant wing against the green jungle canopy adds a delightful dividend to the pursuit of butterflies.

3

Rise of the Angiosperms

DAVID L. MULCAHY

About 125 million years ago, the earth's terrestrial vegetation underwent a dramatic change. At that time, a newly evolved and highly competitive group of plants suddenly rose to dominance. These were the angiosperms, or flowering plants, and they relegated the gymnosperms (most familiar to us as conifers, but also including cycads and the ginkgo) and the ferns to a secondary position. The change was swift, at least in terms of geologic time and, to this day, irreversible. Angiosperms now occupy well over 90 percent of the earth's vegetated surface and, aside from some harvests of marine algae, constitute virtually 100 percent of our agricultural species.

Where did they come from, what were their ancestors, and most intriguing of all, why were the angiosperms able to displace the gymnosperms and the ferns from their position of dominance? We have answers to some of these questions, but to a considerable extent, the rise of the angiosperms remains for modern botanists, as it was for Darwin, an "abominable mystery." Today, in part because flowering plants represent the best hope for increasing agricultural yields and for utilizing the 36 percent of the land surface that is too dry, toxic, or saline to support present crop species, it is more necessary than ever to understand this fascinating group of plants.

On which points are present-day scientists closer to answers than were the Victorians? Thanks to Daniel Axelrod of the University of California at Los Angeles we may at least know something about the environment in which the angiosperms evolved and why their appearance was so sudden. His hypothesis is that the angiosperms arose in arid and mountainous regions where the ponds and lakes that figure so prominently in the formation of fossils would certainly be rare. Without muddy lake bottoms, which are free of decay-causing oxygen and rich in fine-textured sediments that preserve the form and even the tissues of delicate structures, few fossils would be formed. There would be exceptions, of course, and there are some large lakes in mountainous regions. These lakes, however, would never be filled with the silts and clays that often protect fossil deposits in lowland lakes. Thus when mountainous regions are eroded away, the high-elevation lakes and fossils will go with them, and any traces of life they carried will be destroyed. If the Axelrod hypothesis is correct, the first steps in the evolution of the angiosperms may be lost forever.

The hypothesis also suggests something about the population structure of the ancestral angiosperms. Mountainous terrain is likely to be heterogeneous, and if the regions inhabited by the early flowering plants were arid as well, the plants growing there would probably have been limited to particularly favorable sites. Perhaps the first angiosperm populations occupied only the less arid swales and ravines and were separated from one another by hummocks and ridges. This population structure—a series of medium-sized groups separated by short distances—happens to be the one thought to best provide two requirements of rapid evolution: a good supply of genetic variation and a measure of isolation. The genetic variation contained in medium-sized groups may be adequate to produce individuals adapted to local conditions, but even if it is not, an occasional seed or pollen grain from a nearby group could replenish the stockpile. Moderate isolation would allow this occasional input of genetic material while at the same time preventing a flood of material that would disrupt adaptive gene combinations already present within the population.

The Axelrod hypothesis made a major contribution to the understanding of early angiosperm history, but unfortunately it tells little about how these newly evolved plants were able to out-compete and displace their gymnospermous antecedents. Here we must look to the angiosperms themselves. Their competitive superiority has been attributed to many different characteristics. For example, angiosperms generally produce broader leaves than gymnosperms; this enables the flowering plants to intercept more light for photosynthesis. Angiosperms also produce an amazing arsenal of foul-tasting or even toxic substances that presumably protect them from grazing animals and insects. Furthermore, angiosperms, unlike nearly all gymnosperms, possess vessel elements, specialized conducting cells that form highly efficient vascular systems.

Each of these characteristics has been implicated in the adaptive success of the angiosperms, but none is limited to the angiosperms. Ginkgo, a gymnosperm, produces broad leaves; the resins of the conifers clearly discourage even the most voracious herbivores; and *Gnetum*, another gymnosperm, produces vessel elements, as does bracken fern. Moreover, the more primitive angiosperms apparently lacked vessel elements. Perhaps all these adaptive features are more the products of angiosperm success than its cause. At best, they are probably only a part of the answer. For the rest, let us consider another characteristic of the angiosperms: their mode of pollen transport.

The primitive angiosperms were pollinated by insects, as are most of the present-day angiosperms. (There have been some reversions to wind pollination, but these occurred after the essential features of the angiosperms had evolved.) An intuitive presumption that insects are more reliable than the wind as a means of pollen transport is supported by some excellent studies by Robert W. Cruden of the University of Iowa. He has shown that with insect-pollinated species 6,000 pollen grains may be produced for each ovule, a large ratio but one that pales in comparison with the million-to-one ratio found in wind-pollinated species. This disparity suggests that even if insects consume a great quantity of pollen (up to 90 percent according to some estimates), they are still less profligate than the wind.

In addition, because insects fly more or less directly from flower to flower, they are far more likely than the wind to accomplish pollen transport between such moderately isolated subpopulations as the groups of early angiosperms proposed by Axelrod. The wind can carry pollen for many miles, occasionally covering the decks of ships far at sea. As a general pattern, however, pollen grains become widely separated as the wind carries them along, their concentration falling by a factor of 30 million in just 800 feet. This means that beyond several feet from their source, wind-transported pollen grains will be relatively sparse.

If insects are the more dependable means of pollen transport, the issue then becomes one of enticing them to provide this valuable service. The early angiosperms apparently solved the problem by offering insect visitors a meal of pollen, a persuasive inducement since pollen is an extremely rich source of protein. The presentation of a pollen meal was not without problems, however. William Crepet of the University of Connecticut has pointed out that ovules, if they were to receive the gametes of the insect-transported pollen, had to be close to the insect visitors, and ovules are every bit as nutritious as pollen. They are not, however, nearly as abundant. Natural selection apparently favored plants in which ovules were afforded some protection from insect predation, ultimately resulting in the angiosperm ovary, a vessel that completely encloses the ovules. (The word angiosperm means "vessel seed"; gymnosperm means "naked seed.")

At this point it is possible to reconstruct an interesting (and universal) phenomenon. A problem, in this case of the unreliability of the wind as a means of pollen transport, elicits a solution, for example, insect pollination. The solution produces its own set of problems, but unless the solution is no solution at all, these are less severe than the original problem. The secondary problems do, however, necessitate further adjustments in the system: in the evolving angiosperms, ovules had to be protected. Theoretically, this sequence of problems, solutions, and still more problems and solutions continues forever, but in reality, the adjustments probably soon become attenuated to the point of evolutionary insignificance. The important part of this phenomenon is that for many problems, there are several possible solutions. Some of these merely solve the problem at hand, but others may provide capabilities not directly selected for.

The Chinese incorporate an awareness of this duality into their language, using the single word *ji* to mean both "crisis" and "opportunity." In the case of the evolving angiosperms, the unexpected occurred in the third cycle of crisis and opportunity. Enclosing the ovules in protective structures did protect them from insect predation, but it also isolated them from the pollen grains. The solution to this third-level problem lay in the evolution of pollen tubes that could grow from the pollen grains through the intervening protective structures and into the ovules. This solution may have presented the angiosperms with a great opportunity. Consider the fate of the pollen grains of gymnosperms. Carried by the wind, most are lost, regardless of their quality or genetic content. For those that do reach an ovule, the worst is over; developing eggs lie just a few cells away. Even fairly weak genetic types are probably able to complete this final part of the journey. The challenge of this last stage is further reduced by the small number of pollen grains likely to reach any particular ovule, so competition among pollen tubes should be modest.

In contrast, all available evidence suggests that although many of the pollen grains of an insect-pollinated angiosperm are sacrificed to insect nourishment, enormous populations still reach their target, the angiosperm stigma. There they germinate and rapidly produce pollen tubes, which must penetrate the great length of tissue separating them from the ovules. In five to ten hours, for example, the pollen grain of *Petunia* produces a pollen tube whose length is about 900 times the diameter of the original pollen grain. This rate is comparable to a basketball-sized sphere producing a tube about one-fifth of a mile long. Since only a well-balanced and vigorous physiology can sustain such growth, this marathon of pollen tube development presents a challenge to the metabolic system of the pollen. Furthermore, with insect pollination the number of pollen grains reaching the stigma probably greatly exceeds the number of ovules available for fertilization. This creates an intense competition among the pollen tubes, and only the fastest-growing and most vigorous will pass their gametes on to the next generation. Therein lies the surprising opportunity for the angiosperms.

Pollen competition apparently provides a mechanism whereby poor genetic types are eliminated and vigorous types preserved. This idea is not new but neither is it universally accepted. Some evolutionists view the extraordinarily intense competition that occurs among pollen tubes as a real danger point in the life cycle of a flowering plant. In 1932, for example, the geneticist J. B. S. Haldane suggested that a gene that greatly accelerates rates of pollen-tube growth will spread in a population even if it has mildly deleterious effects in other parts of the life cycle. "A higher plant," in Haldane's words, "is clearly at the mercy of its pollen grains." Experimental evidence, however, supports a much more sanguine view. Several investigators, beginning with the Russian geneticist D. V. Ter-Avanesian in 1949, have found that in cotton, wheat, corn, tomatoes, petunias, and carnations, gametes from rapidly growing pollen tubes give rise to vigorous plants. Thus, with apologies to Robert Frost, one might say that "good pollen tubes make good plants." It therefore becomes significant that only the angiosperms have a highly effective method of selecting for rapidly growing pollen tubes.

One of the most powerful aspects of pollen tube selection is that vast numbers of individuals are involved, but the system has a qualitative advantage as well. Remember that within plant embryos, seedlings, and mature individuals, every cell (except the gametes) carries two sets of genes, one inherited from each parent. This diploid state has some advantages, chief among them being the ability to store genetic variation, but it has disadvantages as well. Recessive genes, for example, are expressed only when they are carried in all (here both) sets of genes within a cell. Suppose that, by chance, a rare but highly beneficial recessive gene appears within a population. Because it is so rare, there is little chance that it will occur within both gene sets of a cell. The gene, with its beneficial characteristics, will remain unexpressed. Within pollen, however, each cell is haploid and thus contains only one set of genes. Consequently, any gene beneficial to the pollen, even a very rare one, will be expressed within the pollen grain and favored by natural selection. This beneficial gene will increase within the population and eventually reach levels at which both gene sets in some seedlings will carry it. In these respects, pollen grains are similar to bacteria, organisms that exhibit a truly astonishing ability to respond to selection. We have only to consider the ease with which bacteria have developed resistance to many antibiotics in order to realize how adaptable they are. Microorganisms can produce a new generation as quickly as larger organisms can produce a new cell. Thus they can invade any new habitat or respond to changing conditions with billions of individuals. Also, because microorganisms are haploid, mutations are expressed, and genetic adaptations tested, immediately.

There is another aspect to the potential benefit of a pollen selection system.

George Williams of the State University of New York at Stony Brook and V. C. Wynne-Edwards of the University of Sussex have asked what contributions sexual reproduction makes to the process of adaptation. Granting that it provides new and possibly adaptive genetic combinations, they point out that sexual reproduction also has a negative effect. The sexual process takes two perfectly good genetic types (good enough, that is, to survive to, and perhaps through, the rigors of parenthood) and scrambles them together, risking the loss of the good qualities of both parents. Williams and Wynne-Edwards suggest that unless populations are huge, on the order of tens of millions of individuals, sexual reproduction may cost more than it gains. Here is where pollen tube competition may become important.

In both angiosperms and gymnosperms, for each individual that reaches maturity, large numbers of seedlings are started. However intense this selection among seedlings may be, the angiosperms, as we have seen, possess yet another level of selection. Pollen tube competition may mean that the angiosperms, when compared with the gymnosperms, are relatively free to recombine genetic types, and that the selected products of this recombination are quite certain to possess a reasonably well-balanced, vigorous gene set. The gymnosperms, lacking the opportunity for intensifying pollen tube selection, would be forced to adopt a relatively conservative genetic system, one that preserves adaptive genetic types rather than producing vast numbers of new adaptations.

Intense pollen tube competition, an accidental consequence of the shift from wind to insect pollination, may thus have granted the angiosperms carte blanche for genetic experimentation. This evolutionary plasticity may have led early and often to adaptive features, rare or absent within the gymnosperms, that enabled the flowering plants to reach their present position of dominance in the world's flora.

4

Jewelweed's Sexual Skills

DONALD M. WALLER

From late August through September, the pendent, red-flecked orange flowers of *Impatiens capensis* (*I. biflora* in older manuals) are a familiar sight, gracing streamsides and openings in flood-plain woods of the northeastern and midwestern United States. Rain and dew drops bead up on the light green leaves and sparkle in the sun, hence the common name jewelweed for this succulent-stemmed annual.

The flowers are distinctive, with one of the sepals forming an elongated sac that ends in a curved spur. Inside the spur, a copious supply of nectar awaits bumblebees and hawkmoths with tongues long enough to reach it. Patches of wild jewelweed are also visited by ruby-throated hummingbirds, which dart swiftly from flower to flower, matching the bumblebees' quick, direct movements. These birds, too, have tongues long enough to give them access to the nectar cache. Beige stripes of pollen appear on the backs of the insects or the forehead of the hummingbird. It is a light load, even for the small bee, but to the jewelweed it is all-important. The pretty blossom and its sweet nectar entice the mobile bees and birds into playing matchmaker for the sedentary jewelweeds.

To maximize their chances for spreading pollen to, and receiving pollen from, as many other plants as possible, these flowers have evolved another, less noticeable trait. When the flowers open, they function as males and their five anthers spread sticky pollen abundantly on any visitor. After about a day, however, the flowers drop the anther unit, exposing for the first time the virgin female receptive surface, the stigma. Because each flower progresses from an entirely male phase to an entirely female phase, self-fertilization is prevented and cross-fertilization with another flower enforced. Devices to prevent self-fertilization, such as this simple sequence of sex roles, are common throughout the plant kingdom. Noting these devices, Charles Darwin concluded that "nature abhors perpetual self-fertilization."

In addition to these showy, distinctive flowers, jewelweed produces another kind, minute (one to two millimeters) and inconspicuous. Termed cleistogamous by botanists (Greek for "hidden marriage"), these tiny flowers never open to admit visitors, but instead always self-pollinate. Natural selection has reduced the number and size of their parts to a bare minimum: vestigial petals, no nectar, and only a few pollen grains. These flowers develop from a green, budlike stage directly into an enlarging seed capsule, which may still have its flower envelope attached to its tip like a cap. The cleistogamous flowers grow singly from the axils of leaves on lower branches. The bigger, brighter chasmogamous flowers (from *chasma,* "opening" in Greek) occur in clusters from axils near the top of the plant or the tips of branches. Individual plants usually produce both kinds of flowers, but the cleistogamous flowers are easily overlooked.

Why should a plant with one apparently effective means of producing seeds bother making another kind of flower? Why hasn't natural selection favored one mode of reproduction over the other? What and how great are the dangers posed by self-fertilization for a plant that habitually indulges in it? For the last several years I have striven to answer these questions.

Any effort to understand jewelweed's breeding system requires an understanding of its growth and reproduction. As its name implies, jewelweed is rather common and grows in a variety of habitats. Although most successful in clearings near streams or in marshes with consistently damp soil, jewelweed also manages to invade seasonally dry and sometimes sandy flood plains and even relatively dry and shady sites in some oak forests. Under ideal conditions—well-lighted sites that remain damp throughout the summer—jewelweed grows from a small seed (weighing less than a thousandth of an ounce) in March to a plant up to six feet tall by late August. The plant could not grow so quickly if it took time to construct a woody stem or even the dense pith of many other herbs; instead it builds a hollow, succulent stem, held erect by turgor pressure. In dry or shady locations, jewelweed only grows one to two feet tall and may succumb to a dry spell in July or August.

Despite the diverse weather and site conditions under which jewelweed grows, plants in any given population abandon vegetative growth and begin channeling their energies into flowers and seeds several weeks before they senesce, whenever this occurs. The species' ability to grow and reproduce in a variety of environments is crucial. As an annual without prolonged seed dormancy, jewelweed depends on regularly recurring seed crops to repopulate the stream banks and marsh edges each year. Any trait that insures that at least some seeds are set every year would be favored by natural selection.

Each chasmogamous flower that successfully attracts a pollen-loaded insect or bird develops into a translucent, pendent seed capsule containing three to five seeds. (The seed capsules that develop from cleistogamous flowers are similar except that they usually contain one to three seeds.) When fully ripe, the seeds darken and the capsule becomes sensitive and will burst suddenly at a touch, scattering the seeds. This trait has endeared jewelweed to many children (and not a few adults) who cannot resist the temptation to help the plant along by springing the capsules. It is also responsible for some of jewelweed's other common names, such as snapweed and touch-me-not, as well as its genus name, *Impatiens*.

Once sprung, the seeds are shot up to five feet away. Since they also float, the seeds can disperse long distances down streams and around lakes. By having such capsules, jewelweed avoids having to bribe animal dispersers with fruits to carry seeds away from the parent plant. I have sometimes thought that the exploding habit might have arisen to foil predators. I once watched a finch peck repeatedly at ripe jewelweed capsules, only to have them explode in its face.

The production of both showy, outcrossing flowers and minute, self-fertilizing flowers is not unique to jewelweed nor is it universal in the genus *Impatiens*. The introduced garden species *Impatiens balsamina*, for example, rarely produces cleistogamous flowers. Other common plants that do have selfing flowers include wood sorrel, hog peanut, many violets, and the bush clovers. In some of these species, the same plant produces both kinds of flowers, while in others, individuals produce one kind or the other. In the eighteenth century, the Swedish systematist Carolus Linnaeus observed cleistogamous and chasmogamous flowers on species of *Cistus* (rockrose) and *Salvia* (sage). He further noted that when these plants were transplanted from their warm, native Spain to the colder climate of Sweden, they responded by producing only the closed, self-fertilizing flowers. Botanists since Linnaeus have observed similar responses to stress in many species capable of producing both kinds of flowers. The phenomenon is common enough to have been christened environmental cleistogamy.

The first scientist to give serious consideration to the puzzle posed by cleistogamous flowers was Charles Darwin. In his later years Darwin became especially interested in plant growth and reproduction. In *The Different Forms of Flowers on Plants of the Same Species* (1877), he discussed the incidence of cleistogamous flowers in many species and accurately described their structure. More than once Darwin remarked on how "wonderfully efficient" these flowers were in their economy of pollen and success at setting seed. He further noticed that in species with both cleistogamous and chasmogamous flowers, the chasmogamous flowers are usually medium to large in size and bilaterally symmetric, as in jewelweed. Such flowers effectively exclude many small generalized pollinators and must rely on specialized animal pollinators for fertilization. This, Darwin reasoned, would make the outcrossing flowers vulnerable to complete seed failure if harsh weather conditions or some other adverse event kept pollinators away. Selfing flowers could provide an alternative method of reproduction.

Cleistogamy is more common in annuals than in perennials, for which failure to set seed one season is not fatal. For species that frequently colonize new sites, selfing offers an additional advantage. If it is self-fertile, even a single seedling arriving at a new location can set some seeds and populate the site with its descendants. Thus the value of cleistogamous flowers is clear.

More difficult to understand is why plants don't always rely on self-fertilizing flowers. In attempting to answer this question, Darwin conducted some ten years of experiments, involving thousands of hand pollinations. He demonstrated that for several species of garden flowers (which normally outcrossed), the seedlings derived from continually self-pollinated lines were generally smaller than those resulting from cross-pollinations. (Remarkably, Darwin accomplished all this before the rediscovery of Mendel's laws of inheritance, which provide a genetic explanation for inbreeding depression.) Darwin suggested that cross-fertilization would normally be favored because it produced more vigorous offspring, but that under conditions adverse for pollination, species able to produce seed efficiently by means of specialized self-fertilizing flowers would be at an advantage. Most botanists still accept this explanation.

Does this explain jewelweed's pattern of flowering, too? To find out, I pursued several lines of investigation. At the outset, I thought it important to assess whether different environmental conditions affected jewelweed's breeding system in the same way that they affect species known to exhibit environmental cleistogamy.

I surveyed several natural populations growing in a variety of sites near Princeton, New Jersey. At each site, I took measurements of light intensity, soil moisture, and the proportion of outcrossing flowers. The populations displayed a simple pattern: those growing in areas with both sun and moist soil, such as in openings along streams, produced mostly outcrossing flowers; while those with either dry soil or shaded conditions produced fewer outcrossing flowers. One population growing in a pine plantation with very well-drained soil and continuous shade never produced any outcrossing flowers at all. Jewelweed's growth was so inhibited at this site that the seedlings never exceeded eight inches in height. These plants began to produce their selfing flowers in May before they had even dropped their embryonic leaves. This proved to be prophetic since the entire population died by early July, before any outcrossing flowers could have completed their development. Surprisingly, the seeds produced by cleistogamous flowers alone have been sufficient to maintain jewelweed at this site for years, and the population even appears to be expanding.

Next, I needed to evaluate how much of the variation in flowering was due to jewelweed's ability to respond to different conditions and how much was due to inherent genetic differences among the populations. I grew seedlings derived from a single population under different light and water treatments in a greenhouse. All of the plants devoted a similar fraction of their resources to self-fertilizing flowers. In contrast, as they became larger, the plants invested an increasing fraction in the outcrossed flowers. Sun-grown plants produced a great many outcrossing flowers, while shaded, smaller plants produced only a few. Plants that were both shaded and exposed to drought conditions hardly grew at all and never produced outcrossing flowers, just like the plants in the pine woods. Apparently, a threshold size must be exceeded before a jewelweed plant will invest in outcrossing flowers. Differences in plant size accounted for most—but not all—of the differences in flowering behavior. When I experimentally shaded one group of plants in the field while leaving an adjacent group of the same size exposed to partial sunlight, shading still reduced the number of outcrossing flowers produced.

I also wanted to determine exactly how much "cheaper" seeds produced by selfing flowers were compared with out crossed seeds. The efficiency of these flowers had been evident to all who studied them, but no one had tried to measure precisely how many grams of plant tissue or calories of energy were saved by relying on modest, self-fertile flowers instead of showy, outcrossing flowers with their large corolla and abundant nectar.

To do this, I studied a population in a patch of woods on the Princeton campus that I could visit every other day. Although they covered only a few acres, this patch of oak-hickory forest bordered a stream, which permitted a dense stand of jewelweed to grow. I carefully followed the survival and growth of a number of flowers of each type from their inception as buds through the maturation of ripe seed capsules. The bulk of the outcrossing flowers were initiated in late August and took nearly five weeks to develop a mature capsule. Although most plants initiated a few selfing flowers early in midsummer, these rarely developed unless the plant was injured or suffered drought. Most of the cleistogamous flowers were produced late in the season, in September, after the outcrossing flowers. Since these flowers require only a little more than three weeks to mature seed, they provide a means for setting seed right up until the frosts of October. Clusters that initially contained only showy, outcrossing flowers even reverted to producing selfing flowers, occasionally passing through an intermediate flower type.

By averaging information gathered for more than seven hundred flowers, I could compute the overall probability of a flower of either type surviving from its inception to a ripe seed capsule. For both flower types this was between 50 and 60 percent. Flowers and developing capsules cropped from other, nearby plants provided estimates of how much plant material, or biomass, was invested in flowers or fruits at every stage of development. Using both these biomass figures and may calculations on survival from bud to seed, I then estimated the total production costs for a seed of each type. I found that seeds derived from selfing flowers cost only about two-thirds as much biomass as those from outcrossing flowers.

While I was busy doing these studies in New Jersey, Doug Schemske, now of the University of Chicago, was conducting a similar intensive study of several *Impatiens* populations in Illinois. By carefully tracking the development of the flowers, he discovered that the pollen-release phase lasts about twenty-four hours, while the receptive phase only lasts about four hours. He also found that larger plants invested a greater proportion of their resources in outcrossing flowers and seeds than smaller plants, and that sunlight enhanced this response. Using the currency of calories and including nectar production, Schemske estimated that seeds from the cleistogamous flowers cost one-third to one-half as much as seeds from outcrossing flowers. I was gratified to have confirmation of jewelweed's plastic response and of the necessarily imprecise measurements of seed cost.

My greenhouse work with plants from one population suggests that much of the variation in outcrossing I observed between different jewelweed populations could be caused by plastic response to different environments. Since the investment in outcrossing, rather than selfing, flowers is what varies the most, environmental cleistogamy is really a misnomer for jewelweed. Small plants produce few or no outcrossing flowers, but other stresses can produce this response in larger plants as well. Whenever jewelweed was grazed by deer or when I cut off the ends of its branches in the greenhouse, seeds ripened only from selfing flowers. The response allows jewelweed to mature seeds quickly in the event of damage or drought. This kind of switch-hitter strategy must surely contribute to jewelweed's consistent reproductive success.

But why should large plants be more willing than small ones (speaking teleologically) to pay the high costs of outcrossing? Darwin had demonstrated the superiority of outcrossed plants for several varieties of garden flowers. To determine whether some intrinsic advantage accrues to outcrossed jewelweed offspring, I measured germination and growth of both kinds of seedlings in the greenhouse. There was some superiority for the outcrossed seedlings as a group, but also considerable variability. Since greenhouse conditions do not correspond to growth conditions in nature, I am now comparing the two types of seedlings in field studies.

If there is some consistent benefit of outcrossing, does its cost vary in any predictable way with plant size? There do not appear to be any obvious conditions inimical to pollinators that would decrease the likelihood of a small plant setting seed from its outcrossing flowers. Even in sunlit populations with ample moisture, some small, suppressed individuals produce only the small selfing flowers. Could it be that only large plants can produce enough showy flowers in conspicuous locations to attract the birds and the bees? This seems unlikely. I have not been able to detect any preference among bumblebees for flowers in any particular location, although the placement of outcrossing flowers at the tips of branches could reflect pollinator preferences. The answer may have more to do with the risks of reproduction. Small plants, able to produce only a handful of flowers and seeds, risk not setting any seeds at all if they invest only in outcrossing flowers, which may not be pollinated. Large individuals, with their greater number of both types of flowers, are in a better position to gamble against the odds of no pollinators. For them, the payoff of superior seedlings is worth the risk of a few unpollinated flowers, especially since the seedlings will likely face keen competition the next year in such a favorable habitat.

With so much developmental and physiological flexibility in the face of changing conditions, has jewelweed ever adapted by genetic means to particular kinds of environments? This is an important evolutionary question since if some general purpose genotype performs well in all environments, continued genetic recombination by means of chasmogamy would have little effect. Presumably, some of the new genetic combinations produced by sexual outcrossing are even better adapted to particular site conditions or more resistant to predators and pathogens. In self-fertile species capable of some outcrossing, once a particularly adaptive genetic combination is generated from cross-fertilization, it can be reliably passed along by selfing.

In my greenhouse experiments, I had noticed that New Jersey seeds germinated sooner than seeds from Massachusetts or Wisconsin, suggesting some genetic differences. Robert Simpson, Mary Leck, and V. Thomas Parker of Rider College have now gone further by comparing six New Jersey populations grown together in the greenhouse. They found no genetically based differences in vegetative behavior, but considerable differences in flowering times, which tended to match those observed in the source populations. These inherited differences between the populations indicate that the populations have undergone independent selection, adapting, at least to some extent, to local sites.

Apparently, then, jewelweed has evolved the ability to react tactically in the short term, by means of plastic growth and reproduction, to unpredictable changes in its environment, while simultaneously adapting genetically to certain site conditions. By generating new genetic combinations, outcrossing—a conspicuously expensive reproductive strategy—plays an essential role in this process of adaptation. For jewelweed, as for most species of plants and animals, the benefits of insuring genetic diversity are worth the costs.

5

Toxic Tailings and Tolerant Grass

ROBERT E. COOK

The tailings of abandoned mines hardly seem a fitting place to watch evolution. Vestiges of former excavations, they lie like forgotten scars on the face of the landscape. Rain rapidly drains through the coarse, porous soil, which holds little organic matter and few mineral nutrients. Toxic concentrations of heavy metals—zinc, copper, and lead—contaminate the sediments, rendering the ground water poisonous for most plants. Careful searching across this largely barren surface, however, frequently turns up tufts and large patches of grass that apparently tolerate the toxic metals and thrive under the stress of drought and mineral depletion. The successful adaptation of these rugged plants to derelict mines is a testament to the power of natural selection and provides a fine opportunity to study the process in operation.

More than one hundred years ago, Charles Darwin hinged his theory of evolution on the natural selection of individuals. Through the differential survival and reproduction of variants appearing in each generation, traits that adapt an organism to a particular habitat can increase in frequency in future generations. A century after the publication of *Origin of Species*, British geneticist Anthony Bradshaw and his students at University College of North Wales in Bangor set out to piece together a detailed picture of rapid adaptation among plants to the presence of heavy metals. To do so, they focused on the significance of natural variation.

Bradshaw and his students noticed that although pastures surrounding local mines were rich with species apparently unable to survive on the contaminated mine soils, there were some species growing in both habitats. For example, *Agrostis tenuis,* a wind-pollinated grass found in abundance in the pasture, forms sparse populations on mine soils. Two explanations of this phenomenon seem possible. On the one hand, perhaps certain species, such as *Agrostis,* possess a tolerance mechanism that enables them to extend their range to the excavation site. Alternatively, most of the pasture plants of *Agrostis* may not be able to tolerate heavy metals, and the colonizers of the mine may represent exceptional individuals that have adapted to the metals.

To determine which possibility is correct, researchers have developed an assay to measure the degree of tolerance possessed by an individual plant. They grow selected tillers or seedlings of a test plant in two solutions: one, the control, supplies only the nutrients found in normal soil; the second also contains a toxic concentration of a particular heavy metal. In the second nutrient solution, nontolerant plants do not grow roots, but the roots of tolerant individuals rapidly proliferate. Hence the vigor of the root growth in the toxic medium, compared with the growth in the control medium, provides a reliable index of tolerance. On this scale a rating of zero indicates a lack of tolerance, while a completely tolerant plant is rated ten.

Bradshaw sampled plants of *Agrostis* from a local mine and the surrounding pasture and grew them in the metal solution and the control. The tolerant plants rapidly rooted to form vigorous tillers, but the roots of the pasture plants were stunted and the shoots did not grow. The mine populations were clearly different from those found in the pasture and had evolved heavy metal tolerance since the original excavations. Moreover, although some mining operations date from the Middle Ages, many *Agrostis* populations have been discovered growing on mines less than one hundred years old; this evolution must be remarkably rapid.

Because the tolerant and nontolerant populations of *Agrostis* grow so close together, the explanation for such rapid evolution is more complex than simply quick adaptation to new conditions. Tolerant individuals of *Agrostis* grow well on mine soils in the presence of heavy metals and under harsh habitat conditions. Their offspring share their tolerance, indicating a strong genetic basis for this adaptation. In the pasture, nontolerant *Agrostis* survive nicely, competing vigorously with the wide array of other species in the field. The two populations meet at the boundary of the toxic soil. Since these plants are pollinated by wind, some pollen from each side fertilizes plants on the other side. As a result, some offspring on the mine side of the boundary should inherit nontolerant genes, and seed produced on the pasture side should display some tolerance.

This movement of genes in both directions could be expected to create an array of tolerances among offspring and a continuous transition zone from very tolerant plants well into the mine to very intolerant plants well into the pasture. Surprisingly, there is a very sharp break in the tolerance of the two populations, and the border between them is a clearly defined, sharp line. Although mine and pasture plants are only inches apart at the boundary, the mine individuals are fully tolerant, whereas the neighboring pasture plants fail to grow in the smallest concentrations of heavy metals. How is such a distinct boundary maintained in spite of the flow of genes back and forth between mine and pasture?

A critical tension exists between this movement of homogenizing genes between two different *Agrostis* populations and the selective factors working to maintain particular qualities of the populations. When wind-borne pollen carrying nontolerant genes crosses the border and fertilizes the gametes of tolerant females, the resultant offspring show a range of tolerances. The movement of genes from the pasture to the mine would, therefore, tend to dilute the tolerance level of seedlings. Only fully tolerant individuals survive to reproduce, however. This selective mortality, which eliminates variants, counteracts the dilution and molds a totally tolerant population. The pasture and mine populations evolve distinctive adaptations because selective factors are dominant over the homogenizing influence of foreign genes.

Thus evolution—the continuous sculpting of particular physiologies, morphologies, and behaviors—is accomplished through the differential survival of variants created in reproduction. Plants, because they do not move, lend themselves to observation of the process of evolution. The particular elegance of studying metal tolerance in *Agrostis* lies in our ability to measure variation and watch how it changes with each generation. Natural selection, which may ordinarily run long and slow, is captured in the act, held for inspection by the opposing flow of genes.

Tom McNeilly, a student of Bradshaw's, separated the effects of selection from the flow of genes by observing a population of *Agrostis tenuis* he found growing on a derelict copper mine in Caernarvonshire, Wales. The mine's surface, less than 985 feet across, sits on the floor of a U-shaped, glaciated valley whose steep slopes are covered with sheep pasture. McNeilly suspected that the east-west orientation of the valley, which strongly polarizes the prevailing winds in a westerly direction, might provide an appropriate setting for his work.

Toward the end of the summer, McNeilly collected seeds from a group of marked adult plants on the mine. He germinated thirty seeds and subjected the seedlings to the tolerance test in a nutrient solution with a toxic concentration of copper. The marked adults were then transplanted to a greenhouse where, isolated from the pasture plants, they formed a new batch of seeds exclusively through the exchange of pollen among themselves. These greenhouse seeds were also germinated and their seedlings tested for tolerance. Finally, the marked adults themselves were subjected to the copper solution. McNeilly could then compare the tolerance of adult plants with that of their greenhouse offspring; the tolerance of the same adult plants with that of their mine offspring; and, finally, the tolerance of the mine offspring with that of the greenhouse offspring.

The adult plants displayed a relatively narrow range of tolerance and had an average index between six and seven. A few individuals were as low as five, but considerably more displayed a high tolerance of seven. The seedlings that germinated from the isolated greenhouse seed also displayed an average tolerance between six and seven. As expected in offspring, however, the range of tolerance was greater: one of the seedlings was as low as three, while another, at ten, was completely tolerant. This wider variability among offspring compared with parents is the result of the recombination of genes, through sexual reproduction, into many different sets among the progeny. Because the average indexes of adults and their offspring were identical, we can conclude that the expression of tolerance is controlled by genes. Of more interest is the broad range of tolerance among the progeny compared with the parents. For this new generation to display as adults the same *range* as their parents, the two extremes—low tolerance and high tolerance—would have to be eliminated. This elimination of extremes among offspring is, we believe, the measure of natural selection that maintains the tolerant adaptation among adults.

What about the tendency of gene flow from the pasture to dilute the effects of selection for tolerance? Here all is revealed by the mine offspring, germinated from seeds obtained from the adult plants still growing on the mine. The average tolerance of these seedlings is between four and five, well below that of the parents. The range of tolerance is broad, with three seedlings displaying the highest index of seven and one showing no tolerance at all. Remember, these are the individuals that under natural conditions would disperse across the mine to develop into the adults of the next generation. For the mine population to continue to have an average tolerance of six or seven, most plants with a tolerance below five would have to die. Three-quarters of this offspring population will indeed succumb to toxic concentrations of copper—a very strong selection.

The average tolerance of the mine seedlings is four or five, well below that of greenhouse seedlings at six or seven. This difference must be due to the migration of nontolerant genes: pasture pollen blows onto the mine, pollinates adults, and lowers the average tolerance of their progeny. Adult plants are capable of producing highly tolerant offspring when they interbreed in isolation, but when they grow on the mine in the presence of foreign pasture pollen, the highest tolerance of their offspring is seven. Thus, the population can never achieve total tolerance as long as mine individuals continue to cross with pasture plants. The force of natural selection is tempered by the flow of genes.

The polarity of the pollen-bearing wind in the valley where McNeilly worked permitted a field test of his findings. He marked off two transects across the boundary from the mine soils into the surrounding pasture: one was set across the prevailing wind, and the other ran downwind from the mine into the pasture. At selected locations McNeilly sampled plants and their seed, tested them for tolerance, and compared the average index of parent and offspring to distinguish the effects of gene flow.

As expected, adult plants on the mine side of the first transect, running crosswind, are quite tolerant, with an index of six. The seedlings they produce have a considerably lower tolerance of about four. The difference in the tolerances of the adults and their seedlings provides a measure of the movement of nontolerant genes onto the mine soils. As McNeilly sampled closer and closer to the boundary between the mine and the pasture, the tolerance of both adults and offspring gradually dropped, to about four and three, respectively. About sixty-five feet into the pasture, *Agrostis* seedlings had a tolerance index of just over one—identical to that of the parent plants and close to that of completely nontolerant *Agrostis*.

The second transect presented a very different picture. Close to the boundary on the mine side, tolerance among adults was high, at five. The tolerance of the seedlings of these plants was only a single index point lower, possibly indicating the slight effects of pasture pollen from the upwind side of the mine. Surprisingly, this situation remained the same for more than 165 feet downwind into the pasture. Adult plants of *Agrostis tenuis* growing on pasture soil displayed a high index of tolerance and so did their progeny. Clearly there was a great infiltration of tolerant genes into the pasture population. At 230 feet from the boundary, however, the entire situation was reversed. There parent plants had a tolerance index of three while that of their seedlings was much higher, between four and five. At the last station sampled, almost 500 feet from the mine, progeny still displayed a relatively high tolerance, between three and four, although their parents, with an index of two, were nearly nontolerant. This phenomenon can only be explained by two processes: the constant migration of tolerant genes from the mine, which continues to produce higher tolerance in seedlings each generation, combined with strong selection *against* metal-tolerant seedlings in nontolerant pasture populations. Tolerant genes, which adapt plants to growth in soils with high copper content, are acquired at a cost: such individuals cannot survive in normal pasture populations.

We do not fully understand the biology behind this constraint. It may involve the tolerance mechanism, which seems to enable the plant to sequester the toxic metals in an innocuous complex in the cell walls. We do know that without neighbors, tolerant plants are perfectly capable of growth on pasture soils. Competition with normal plants completely suppresses them in pastures, however, and garden experiments with mixtures of the two inevitably lead to the dominance of the pasture plants. The tolerant individuals seem unable to corner their share of light, nutrients, and water. Plants found on mines are generally different from nontolerant specimens in many ways, however, so this competitive inferiority may derive from biological characters other than the tolerance mechanism. One study of zinc-tolerant *Anthoxanthum odoratum,* sweet vernal grass, found that such individuals are smaller in stature than nontolerant plants, have smaller leaves, and are able to grow more effectively in soils with low levels of nutrients. They also flower earlier in the season and self-pollinate at a high frequency. Such reproductive characteristics tend to isolate the mine population from the flow of foreign pollen, which would otherwise lower the average tolerance among progeny. This incipient reproductive isolation of an interbreeding group of plants is the first step on the path to speciation. Perhaps the evolution of what are now two populations of *Agrostis tenuis* will in time result in the formation of two independent species.

The metal tolerance story has a pleasing epilogue in the real world of land reclamation. Bradshaw has turned his attention to the use of tolerant populations in the revegetation of contaminated soils. Derelict mines are more than an eyesore; they are also subject to severe gully and sheet erosion from surface drainage and wind. The flooding of neighboring streams and the contamination of ground water redistributes toxic spoils into surrounding pasture and agricultural land, sterilizing the vegetation and endangering grazing livestock.

In 1969 Bradshaw and his students began experiments at a number of sites. Preliminary testing of the soils revealed that phosphorus is a major nutrient limiting the establishment of plants over the whole surface. At each site they laid out a set of small half-yard plots differing in topography and applied a series of fertilizer treatments to facilitate initial growth of seedlings. They harvested seeds of tolerant populations of *Agrostis tenuis* and several other tolerant species from contaminated mines and sowed them at high densities, along with commercial, nontolerant seeds that served as controls. When growth was assessed at the end of a year, tolerant plants consistently outyielded nontolerant plants. Fertilizer applications were essential, however, and the pH of the soil determined which tolerant species were most successful: *A. tenuis,* for example, prefers strongly acidic conditions.

These experimental populations of tolerant plants proved remarkably persistent, maintaining vigorous growth for more than nine years. The conventional process of reclamation involves the surface addition of a layer of sewage sludge or topsoil, followed by the sowing of commercial varieties of grass. Bradshaw has calculated that the cost of using metal-tolerant populations with appropriate fertilizer treatments may be less than one-sixth the cost of the traditional method.

Like the best of biomedical science, Bradshaw's study of heavy metal tolerance is a paradigm of rigorous research and a prescription for increasing our well-being. But by focusing on the meaning of variation, evolutionary biology differs from medical biochemistry and physiology, where variation is often considered an unfortunate experimental error. Biomedical investigators tend to concentrate on the commonness of events, on the average pattern seen beneath the glass. Students of ecology and evolution, by contrast, are concerned with the rich pageant of diversity in nature. Far from an annoyance to be minimized by meticulous control, natural variation is recognized as essential to answering the question, Why are there so many kinds of living things?

The answer can have practical applications. Today, the National Seed Development Organization, for example, is selling cultivars of metal-tolerant *Agrostis tenuis* ready for use in reclamation. In a time when each dollar of the U.S. government's budget for pure science is subjected to intense scrutiny and each week reveals a new chemical dump in someone's backyard, we may look to Bradshaw's work for inspiration. Creative, rigorous science need not always be expensive, inexplicably abstract, or irrelevant.

PART 2

THE EVOLUTION OF SPECIES

Biologists love to remind one another, usually with a wry grin, that Darwin never really addressed the "origin of species" in his most famous of books, *On The Origin of Species by Means of Natural Selection*. Darwin's main concern was to establish the very notion of "descent with modification" as the explanation of organic diversity in the minds of his contemporaries. The rival theory, Genesis-inspired creationism, held that species are fixed immutable entities that have undergone no change in the six thousand-odd years thought to have elapsed since their first appearance on earth. Darwin had to combat the notion of species fixity as the very antithesis of evolution, and in so doing, he nearly succeeded in getting rid of the entire concept of species.

Yet the organic world is obviously divided up into numerous distinct reproductive communities. Most organisms reproduce sexually at least in some phase of their life cycles; naturalists have observed for hundreds of years—and, indeed, the savants who wrote Genesis clearly also saw—that sexually reproducing organisms "breed true." Like begets like. Species came to be seen as reproductive communities whose members mate with each other, but not, as a rule, with members of other reproductive units.

In the late 1930s and early 1940s, Theodosius Dobzhansky, a geneticist at Columbia University, and Ernst Mayr, an ornithologically inclined systematist at the American Museum of Natural History, picked up a strand of biological thought that had been rather neglected ever since Darwin won the world over to evolution. Pointing out that there need be no theory of their origins if species cannot be said to exist in nature, Mayr in particular strove mightily to establish the notion of allopatric, or geographic, speciation as the predominant, perhaps even the sole, mechanism of species' origins. The theory sees new species arising from old when a portion of the ancestral species becomes physically isolated. Prevented from interbreeding by geographic barriers, breeding systems and other aspects of the organisms' adaptive systems may become altered from the condition of the ancestral species. Should that drift go far enough before contact is reestablished (or so the theory goes) the ability to interbreed may be diminished, or even lost altogether. Thus new reproductive communities tend to bud off from old ones as a consequence of changing patterns of geographic distribution—changes often induced by the physical transformation of the face of the earth.

Theories of speciation form the central theme of the three essays of part 2. Peter and Nicola Grant present a review of allopatric speciation theory, applying it to Darwin's finches on the Galapagos Islands. The Grants use their data to test the notion that all speciation must be allopatric. They also ask if divergence typically proceeds far enough in isolation to produce speciation, or whether natural selection completes the job only after two closely related, fledgling species come back into contact with one another—a phenomenon ("reinforcement") first championed by Dobzhansky.

Kenneth Kaneshiro and Alan Ohta continue the theme of island evolution in their examination of the incredible diversification of Hawaiian pomace flies—members of the family Drosophilidae more familiar to most of us as "fruit flies." Species are reproductive communities first and foremost, so it is utterly appropriate that Kaneshiro and Ohta have concentrated on courtship and egg-laying behavior in these flies. As the Grants pointed out, island situations have long appealed to naturalists probably because the evolutionary systems in these "natural experiments" are a bit simpler than the more complex and spread-out situations typical of continental mainlands. With the Hawaiian drosophilids, though, nature meets the laboratory in a very graphic sense: most of the early breakthroughs in basic genetics and, later, in evolutionary genetics, came from experiments with fruit flies. They are much beloved of geneticists because of their rapid breeding, ease of handling—and the discovery of giant banded chromosomes in their salivary glands, which aided the early mapping of their genes. With so much known of their genetical systems in general, the Hawaiian drosophilids were a natural to attract the rapt attention of evolutionists.

The final essay, by myself and my wife Michelle, was the first exposition of the theory of "punctuated equilibria" written for a general audience. (The technical paper that coined the phrase "punctuated equilibria" was published, by me and Stephen Jay Gould, in the same year—1972). Some of my professional colleagues tell me that they first really understood what punctuated equilibria is all about by reading the *Natural History* piece.

Punctuated equilibria involves the application of allopatric speciation theory to the fossil record. The article recounts the marked evolutionary stability of a particular group of trilobites. Rather than relying on the model of gradual, adaptive transformation of organic form through time (the picture of adaptation in geological time painted by Darwin and most who came after him), geographic speciation can produce new reproductive communities, differing somewhat in the anatomical properties of their member organisms, in a few thousand years. Once established, a species can live on for millions of years, often with little additional adaptive modification. The pattern in the rock record is of sudden change, but the explanation sees geographic speciation taking some five to fifty thousand years—or so the theory goes.

6

The Origin of a Species

PETER R. GRANT
and
NICOLA GRANT

Between 8000 and 9000 species of birds inhabit the world today. Many more species once existed but have since gone extinct. How can we account for the sheer number of species? The theory of natural selection provides a powerful explanation of how organisms change over time, but one hundred years after Darwin, biologists are still struggling to understand the details of the central evolutionary process—the development of two species from one.

The basic unresolved question is whether speciation in birds occurs primarily (or even only) when populations are physically isolated from one another or whether speciation can also take place when populations are adjacent to one another and thus in contact. According to the first view, any of a number of events, ranging from movements of the earth's crust to a few birds blown off course by a storm, may split a population into two or more geographically separate ones. Over time, these isolated populations adapt to their respective environments and become so different—genetically, behaviorally, ecologically—that they never or rarely interbreed even if chance should bring them together again. This kind of speciation is known as allopatric speciation.

The chief alternative to this view is the idea that speciation may result from either the divergence of two adjacent populations, as each adapts to different ecological conditions in its environment, or the divergence of two segments of a single population, as each adapts to different niches. The first of these two possibilities is often referred to as parapatric, or clinal, speciation; the second, sympatric. Still another possibility is that speciation may involve divergence both when the populations are isolated and when they are in contact.

Regardless of the geographical mode of speciation, there are two kinds of isolating mechanisms that prevent extensive interbreeding between the species. First, individuals may be genetically capable of interbreeding but not do so because they have evolved incompatible courtship displays or songs. Alternatively, individuals of different species may try to breed together, but their mating success is reduced in several ways: the sperm may fail to fertilize the egg; if fertilized, the egg may die; or if offspring are produced, they may be sickly, poorly adapted to the environment, or sterile.

Because the changes that culminate in speciation generally happen relatively slowly, direct field observation of the process is virtually impossible. Instead, ornithologists wanting to identify how speciation occurs must infer the process from present-day results. By comparing the morphology, behavior, and distribution of closely related birds, by examining their genomes biochemically, and by integrating the results with information from studies of other organisms, ornithologists hope to provide a coherent explanation of how bird species proliferate.

The traces of past speciation are easier to follow for some species than for others. The fourteen Darwin's finches—thirteen species on the Galapagos Islands and one on Cocos Island to the northeast—are suitable for such studies for two reasons. In the first place, Darwin's finches are closely related. They are more similar to each other than any of them is to any bird living in the areas most likely to have supplied the ancestral stock: the mainland of South or Central America (although they bear an intriguing resemblance to some West Indies finches). Their closeness supports the idea that they have all been derived from a common ancestor.

Second, the Galapagos Islands are only three to five million years old. The finch species can thus be no older than this, and because they are biochemically similar to each other, they are probably much younger. Since their speciation has occurred in a short period of time, it is relatively easy to interpret. In contrast, in very old lineages many of the links between ancestral and living species have disappeared, which means that unraveling the history of these birds can be nearly impossible.

The first comprehensive reconstruction of the history of Darwin's finches was attempted about forty years ago by the late British ornithologist David Lack. Building on the work of Charles Darwin, Harry Swarth, and Erwin Stresemann, and on the neo-Darwinian synthesis of evolutionary ideas that crystallized in the early 1940s, Lack offered an explanation that had enormous influence, both on how subsequent ornithologists thought about speciation in these finches and on ideas about speciation in general. Now, however, more is known about the finches and their island habitats, and for the past ten years, we have been attempting to take a fresh look at this group of birds to see if they did indeed always evolve as Lack envisioned.

Lack proposed a four-stage process for the evolution of Darwin's finches. In the first stage, some individuals of a mainland finch species either flew directly to the Galapagos and successfully colonized one of the islands or flew first to Cocos Island, their descendants eventually dispersing to the Galapagos. The colonists must have crossed the water because the Galapagos, which straddle the equator approximately six hundred miles west of the South American continental margin, have never been connected to the mainland.

In the second stage, some members of the original colonizing population dispersed to one or more different islands, where they established new populations. Some evolutionary change took place in these derived populations, but Lack, believing that volcanic islands were all ecologically similar to one another, thought the change would not have been great.

The next stage began when finches from one of the derived populations flew to the island holding the original population. The outcome of this secondary contact between birds from the two populations would have depended on how different they were in appearance and behavior. If the derived population had changed very little, interbreeding would have been frequent, and the two populations would have fused. If, however, the populations had become moderately different, the birds would have tended to discriminate, selecting mates similar to themselves. When interbreeding did occur (presumably between the most similar members of the two populations), it would have yielded relatively unfit offspring. Often, the maternal and paternal genes would have been incompatible, resulting in substandard growth, development, and reproduction. Even if the offspring were healthy, Lack believed that since they represented a mixture of the two parental types, they would have had none of the particular advantages of either. Consequently, when competing for food with nonhybrids, they would have been at a disadvantage. Either way, natural selection would have acted against the interbreeding individuals, thereby tending to enhance the differences between the populations and ultimately creating full reproductive isolation between them. In this way, two species would have formed.

The fourth and last stage involved multiple repetitions of stages two and three, culminating in the existence of fourteen species, more if some are extinct. The whole process, a classic case of adaptive radiation, produced the strikingly distinctive birds familiar to numerous biology students. Most of the changes that took place involved beak morphology and function. In one case, tool using developed, enabling the birds (woodpecker finches) to extract arthropods from crevices and holes in branches with the aid of a twig, spine, or leaf stalk. Other birds (large ground finches) evolved extraordinarily large beaks, capable of cracking large, hard seeds. Sharp-beaked ground finches developed the habit of piercing the skin of sea birds and drinking their blood.

In Lack's model, speciation was initiated through morphological divergence in isolation, or allopatry (stage two), and completed when the two populations came together (sympatry, stage three). Is this model, which we call the partial-allopatric model, the best one, according to present-day knowledge? A possible alternative model also involves four stages but with an important difference. Unlike Lack's, this second, simpler model suggests that so much adaptive change took place while the populations were geographically isolated, that at the time of secondary contact neither interbreeding nor competition for food occurred. As a result, there was no selection for divergence, and the derived population fitted into the new island environment without influencing, or being influenced by, the original population. According to this scheme, the complete allopatric model, two species formed from one entirely in isolation from each other, and reinforcement of isolating mechanisms during the period of secondary contact played no part in the process.

We have much evidence to support Lack's model, but the possibility of speciation occurring along the lines of the second must by no means be excluded. In 1961, Robert Bowman, of San Francisco State University, pointed out that the island habitats differ far more than Lack believed. Dispersing from one island to another, therefore, the finches would have encountered a different food supply, and over the next several generations, the birds could have changed considerably as they adapted to the new foods.

In any comparison of the two models, the crucial question is whether evolutionary changes that occurred in isolation were sufficient to account for the differences between species coexisting today. Our field studies have shown that in one case, at least, they have not been enough. The large central island of Santa Cruz is the home of several finch species, including *Geospiza fortis*, the medium ground finch, and *G. fuliginosa*, the small ground finch; as far as we know, these two species never interbreed on Santa Cruz. On the nearby satellite island of Daphne Major, only the medium ground finch has a breeding population. Small ground finches occasionally fly from Santa Cruz to Daphne Major, however, and some of those few that stay hybridize with the medium ground finch. What is it, in this case, that predisposes individuals belonging to two different species to interbreed? The answer is apparently that the medium ground finch tends to be smaller on Daphne Major than elsewhere, hence a little closer in size to the small ground finch.

The important point with regard to the two models is that the differences between the small and the medium ground finches that live together on Santa Cruz cannot be accounted for entirely by changes they underwent while they were separated. If the adaptations during geographical isolation (allopatry) had been sufficient, the medium ground finches on Daphne Major would be just as different from the small ground finches as those on Santa Cruz are. Instead, they are morphologically intermediate between the two Santa Cruz species.

These observations support the claim of Lack's partial-allopatric model that full speciation involves, during secondary contact, an enhancement of the differences between previously separated populations. Further support has come from experiments recently performed by Laurene Ratcliffe while at McGill University. These experiments were designed to test how well birds of each of the six species of ground finches (all in the genus *Geospiza*) discriminate between members of their own species and members of a different species when choosing a mate. All the ground finches engage in similar courtship displays and have identical plumage color. These characteristics, therefore, probably do not help the birds tell each other apart. Beak morphology and song, though, do vary from species to species, so Ratcliffe concentrated on these features.

The goal of the experiments was to see whether species that naturally live together are any more apt to show discrimination than are species that naturally live apart and therefore normally have no need to make discriminations. If they are, this would provide additional support for Lack's model. To test the birds' ability to discriminate on the basis of morphology, Ratcliffe presented different species with lifelike museum specimens of both their own species and other species; to study the birds' reactions to songs, she performed similar experiments using tape recordings of various songs. In both types of experiments, Ratcliffe found that on islands where two species coexist, the birds do indeed discriminate, showing a stronger response to the specimens and taped songs that represent their own species.

She then simulated stage three of Lack's model by experimentally confronting finches on one island with specimens or songs obtained on another island. When male medium ground finches on Daphne Major, for example, were presented simultaneously with a female medium ground finch from Daphne Major and a female small ground finch from Santa Cruz, they courted them equally. The same lack of discrimination was observed in most of the tests conducted with finches from other islands. Birds generally showed a preference for their own species only when the others were very different in size and shape. The failure of many geographically separate species to tell one another apart when experimentally brought into contact argues against the complete allopatric model and supports Lack's model.

According to Lack, natural selection works against individuals that make mistakes in mate selection. We have tried to test this idea by following the fates of hybrids. Unfortunately, so far, we have not been able to determine if interbreeding between the two ground finches on Daphne Major is penalized. Most hybrids known to us have died within the first year of life, but so have most purebred medium ground finches born at the same time. Two hybrid males did survive for three years, but they failed to breed because those female medium ground finches breeding for the first time in these years preferred to mate with particularly large males of their own species. Male hybrids may thus be at a disadvantage because of their small size, but we are not yet certain. Since none of the hybrids has succeeded in mating, we also still do not know if major barriers to gene exchange between the species—such as hybrid infertility—exist.

Apart from any obstacles to reproduction, hybrids would probably have difficulty surviving in a habitat with limited food supply. Lack stressed the role of natural selection in promoting divergence—and thus minimizing competition—between incipient species, particularly in feeding attributes. Unfortunately, this process can only be investigated indirectly. If competition for food has occurred, the differences in beaks (the external feature most intimately related to feeding habits) between two coexisting species should be greater than the differences between beaks of geographically separate populations of those species. If most divergence took place while the populations were still separated, then competition could not have been the promoting factor and the difference between species should be the same, on average, in allopatric and sympatric populations.

Ideally, to test these expectations we would identify the particular geographically isolated populations that have given rise to coexisting species and restrict our comparisons to them. Perhaps one day, with the aid of genetic markers, this identification will be possible, but at present it cannot be done. Neither, given the scattered positions of the Galapagos islands, can we trace individual pathways of colonization. Therefore, we must make our comparisons with large clusters of coexisting and isolated populations, without regard to their possible origins.

Mathematically, the six species of ground finches can form fifteen unique pairs. Thirteen of these possible combinations are known to exist on one or more of the islands. We first studied how much the species in each of these natural pairs differed, on average, in beak size and shape. Next, we artificially combined pop-ulations of the same species on separate islands and examined how much the beaks of these artificial pairs differed on average. Comparing the two groups, we found that in eleven out of thirteen cases the pairs of coexisting species differed more than did the pairs of geographically isolated populations of the same species. Such a high proportion cannot have arisen by chance and strongly suggests that on islands where different species currently coexist, competition for food existed in the past and was reduced by natural selection.

In summary, there are strong reasons for believing that speciation in Darwin's finches sometimes happened in the way proposed by Lack's model: isolation alone does not always seem to produce sufficient differentiation for complete speciation; enhancement of the differences during a period of secondary contact seems to be necessary in some cases. But can the lessons learned in the Galapagos be applied to other birds and other places? After all, most birds inhabit vast continental areas, not archipelagoes.

Proponents of one view claim that Darwin's finches do indeed offer us a microcosm of speciation in general but that the process is harder to detect in large and seemingly more uniform continental areas. Carefully examined, the continents turn out to consist of a patchwork of habitats, each suitable for some kinds of birds and not others. Isolated by surrounding areas of dissimilar habitats, these patches are analogous to islands. Cloud forests, for example, occur on mountains at high altitudes and are widely separated from each other by different kinds of forest at lower elevations. Furthermore, the particular patchwork of "islands" we observe today has not always existed. Past fluctuations in climate, principally during the recent ice age, kept the number, sizes, and conditions of the islands, as well as their degree of isolation, in a state of flux.

Birds can fly, of course, and should be able to traverse habitat barriers relatively easily, especially over continental areas, where they can rest on the way. Our fascination with the powers of migrating species, however, may lead us to exag-gerate bird mobility. Most species of birds live in the tropics, and many of these do not migrate. Nevertheless, the undeniable dispersal powers of birds force us to consider still other possible modes of speciation, those not requiring geographical isolation of populations.

Both of the alternatives—parapatric (speciation of adjacent populations) and sympatric (speciation within a single population)—are plausible in theory, but only under special conditions. There is good evidence for nongeographical specia-tion in plants and insects, but no strong case has yet been made for birds. This may, however, simply be because few searches have been made. If birds do sometimes speciate parapatrically or sympatrically, the evidence is only going to be revealed by thorough field studies.

Recently, we worked with Rosemary Grant on her investigation of what appeared to be incipient sympatric speciation in another of Darwin's finches. In 1978, an interesting subdivision was discovered in the population of large cactus ground finches *(Geospiza conirostris)* on the island of Genovesa. Individual males sang either of two song types, which in itself is not unusual among Darwin's finches; what was unusual was that males with one of the song types had beaks about 6 percent longer, on average, than males with the other type. Their offspring differed in an unusual way, too. Nestlings had either pink or yellow beaks, and we found that the frequency of nestlings with yellow beaks was distinctly higher in broods fathered by one group of males than in those fathered by the other. A similar polymorphism in domestic fowl is known to be genetically controlled. If the differences in the cactus ground finch are also heritable, the possibility exists that the subdivision we observed in the population is an early stage of sympatric speciation.

The two groups of males fed in different ways in the dry, postbreeding season, when food was relatively scarce and many birds, mostly immatures, died. This suggested to us that a degree of ecological isolation had been reached between two segments of the population, again in accordance with the model of sympatric speciation. We have also established that sons sing the songs of their fathers, a prerequisite for discrimination in mate selection. The daughters, however, have so far paired equally with both types of males, regardless of their fathers' song type. This seems to rule out the presence of even partial reproductive isolation between the two population segments, at least under the conditions prevailing during our study. Furthermore, the difference in beak length between the two groups of males was not replicated in the next generation.

These findings raise more questions than they answer about the origins of the differences between the two types of male cactus finches and their offspring. There may well be tendencies toward both fusion and fission at work in the population, and perhaps they alternate in response to the pronounced climatic fluctuations in the Galapagos. Whether, given the right circumstances, the fission tendency could lead to full sympatric speciation, we cannot yet say. Only one sure conclusion can be drawn: much remains to be learned about population structure, mating patterns, gene flow, and natural selection in general before the final word can be spoken about speciation in Darwin's finches or indeed in any other group of birds.

7

The Flies Fan Out

KENNETH Y. KANESHIRO
and
ALAN T. OHTA

Hawaii has more than 500 described species of endemic pomace flies; approximately 200 additional species are in our collection at the University of Hawaii but have yet to be described and classified, and more than 200 species are estimated to await discovery. This large complement of species seems to be descended from a single or, at the most, two original colonist species. Pomace flies today occupy habitats as disparate as morning glory flowers near sea level and slime fluxes of *Myoporum* trees growing on the slopes of Mauna Kea at elevations of up to 7000 feet. They have been found in desertlike environments where the soil is powdery dry, in rain forests with lush tree-fern jungles, and in swampland perpetually shadowed by rain clouds and with vegetation that is burdened with dripping, moss-laden branches.

A team of evolutionary biologists has spent nearly two decades of intensive interdisciplinary research in an effort to understand the mechanisms that have enabled these flies to adapt to the diverse Hawaiian environments. These studies have included the fields of behavior, biochemistry, developmental biology, ecology, external and internal morphology, genetics, physiology, and systematics. We have focused our research on two behavioral traits—oviposition and courtship—that may be important factors governing the rate of evolution as well as the tremendous diversity in the Hawaiian Drosophilidae. For this study we have used an especially striking group of large flies loosely called the picture-wings after the black-and-white wing patterns they display.

An explanation of species evolution in Hawaii must begin with the original colonization events. When a new population is founded on an island, the colonizers may consist of a few individuals or even a single fertilized female. The founders are not select individuals from the ancestral population; flies, for example, are randomly picked up by prevailing winds, high jet streams, and tropical storms, and carried from one island to another. Under such conditions, only a portion of the total gene pool of the ancestral stock is represented by the founders. Thus some of the genetic variability present in the ancestral stock is lost.

The founders, having survived relocation to another region or island, must find a suitable habitat in which to live and breed. During the early stages of colonization, the size of the newly established population is small. The forces of natural selection—competitors and predators—will be subdued, permitting the survival of genetic combinations that were restricted by natural selection in the ancestral population. These conditions allow for new genetic combinations upon which natural selection can later act, enabling differentiations to occur in the colonizers.

Some of the first changes among the early colonizers are in courtship and mating behavior patterns. Such behaviors evolved to allow mate recognition within single interbreeding populations and play an important role in maintaining the integrity of each population's gene pool. Courtship behaviors are important in assuring, with high probability, that the mating between two individuals will result in the best genetic combination for the survival of their offspring.

The most visible indicators of the importance of courtship in the speciation of Hawaiian Drosophilidae are the elaborate behaviors and morphological structures manifested by the males of this group. *Drosophila clavisetae* exemplifies the intricacy of the courtship dance performed by Hawaiian pomace fly males and the morphological changes that have occurred in response to sexual selection. Biologist Elwood Zimmerman described it in the following way: "Few persons are so fortunate to see such wondrous things as [the] male *Drosophila* which has evolved huge scent-dispersing brushes at the end of his tail which he curls over his head and shakes at his lady love to overwhelm her with a shower of aphrodisiac perfume."

Before mating can occur, male pomace flies must perform other chores as part of their lek behavior, which was discovered by Herman Spieth of the University of California at Davis. A lek is an area made up of several adjoining territories from which males display to attract receptive females. (Lekking was first described in birds such as the sage grouse of North America and the bird of paradise of Papua New Guinea.) At a lek of Hawaiian pomace flies, males of a given species establish and defend territories on leaves, stems, or limbs of plants.

The males of each species have developed unique patterns of aggressive behavior, which may in turn be responsible for some of their ornate secondary sexual characteristics. For example, the males of *Drosophila heteroneura* have a mallet-shaped head that has presumably evolved in response to the way in which male-male aggression takes place. Males of this species will stand head to head and, like two rams, butt and shove until one turns or flies away. Another species, *D. nigribasis,* puts on a more delicate performance in defense of its territory. The males stand with their legs and wings fully extended, giving the appearance of standing on tiptoes with wings stretched to the sky. In this position, and facing head to head, the males rock slowly from side to side, alternately touching their wing tips until one leaves the scene. Males that succeed in establishing themselves in a territory will then advertise their presence—by waving their wings or, in some cases, through pheromones—to attract sexually receptive females.

Kenneth Kaneshiro has hypothesized that during the first few generations of new colonizations, when the population size is very small, interspecies interactions are either rare or nonexistent, allowing a relaxation of the normal mate-recognition system. Under these conditions females that are less discriminating in mate selection will be those contributing the most progeny, carrying their genes to the next generation. Females that are too discriminating will be less likely to mate, resulting in a shift in the genetic basis of the mate-recognition system and thus in a simplification of the mating ritual.

Because the Hawaiian fly populations are the products of founder colonizations, the initial changes that did occur in the mate-recognition system were strictly random. Thus, if two populations were separately formed by independent founder events from a common ancestral species, the resultant simplications in the courtship behavior of these populations may have differed to some extent.

Our data indicate that such behavioral changes have occurred in derived populations. We have observed that males of a newly derived population are not able to satisfy the courtship requirements of females from an ancestral population, probably due to the more simplified courtship display of the males. On the other hand, females from the derived population readily accept the courtship rituals of ancestral males, presumably because the males display a courtship repertoire that more than satisfies the females' simplified tastes. This one-sided mate preference between two populations can influence the direction of evolution among closely related species or among isolated populations of the same species. We are attempting to verify our hypothesis by finding the specific changes in courtship behavior that occur within isolated populations of a single species.

The usual forces of natural selection regain their importance in the evolutionary process as the size of a colonizing population increases and other related species are encountered. The increase in interactions with other species results in an increase in the complexity of the courtship rituals to insure species recognition. Thus, we would expect that, in general, older species complexes would tend to have more elaborate courtship patterns than the more recently derived complexes. This is indeed the trend among the picture-winged species group of Hawaiian *Drosophila.* Chance events, then, are responsible for a rapid rate of evolution over a relatively short period of time, while selective forces act at a reduced rate over a longer period and result in the diversity among the species we see today.

The ovipositional behavior of the adult female has become the second major focus of our project to determine the forces controlling evolution in the pomace flies. Individuals of the insect order Diptera (true flies) undergo complete meta-morphosis and have four distinct life cycle stages—egg, larval, pupal, and adult. In most species, the feeding sites of the adults are more diverse than those of the larvae. This is because larvae are much less motile and so are confined to the substrate in which the eggs are laid. The larval feeding site is therefore the same as the breeding site of the species.

The early pioneering work of William Heed of the University of Arizona and the later work of Steven Montgomery of the University of Hawaii have provided us with an overview of the variety of breeding sites to which these flies have adapted to avoid competition, predation, and desiccation, while maximizing the availability of resources. The decaying bark, leaves, stems, and roots of the native Hawaiian plants are the primary breeding sites for the majority of pomace fly species. Others include ferns, fungi, flowers, fruits, and sap exudates (slime fluxes) produced by a few species of native trees. One genus of flies *(Titanochaeta)* has even adopted a parasitic life form, its larvae feeding on the eggs of spiders.

One of the best examples of an adaptation into separate ecological niches, resulting in the avoidance of competition, is the relationship between two very closely related picture-winged species, *Drosophila silvarentis* and *D. heedi.* The females of these species are morphologically indistinguishable from one another, although the males can be recognized by differences in the bristle patterns of the forelegs. Moreover, these species have the same or overlapping ranges over parts of the Big Island of Hawaii. In the dry forest area of the saddle between two large volcanoes, Mauna Kea and Mauna Loa, where virtually no standing water exists and precipitation is quickly taken up or evaporated, both species feed on the slime flux exuded by the native *Myoporum* trees.

Eggs and larvae are also present in the fluxes. We removed these fluxes and brought them to our laboratory where the eggs and larvae present in them were allowed to develop to the adult stage. Surprisingly, all the adults were *D. silvarentis;* the location of the breeding site of *D. heedi* remained a mystery. On a subsequent field trip, we discovered a dark patch of wet soil. This soil patch was moistened by a slime flux dripping from a horizontal branch of a *Myoporum* tree high above, and in it were *D. heedi* larvae and pupae. The ovipositional behavior of one of these species has evolved to allow a partitioning of scarce resources.

Two other picture-winged species, *D. grimshawi* and *D. crucigera*, possess properties that make them ideal for the study of ovipositional behavior. These are the only two drosophilid species of the picture-winged group not endemic to a single island. *D. grimshawi* inhabits all the major islands except Hawaii, and *D. crucigera* inhabits the islands of Kauai and Oahu.

Populations of *D. grimshawi* found on Maui, Molokai, and Lanai and *D. crucigera* throughout its range are the only generalists in the picture-winged group. The majority of picture-winged species are highly host-specific, utilizing only one to five plant families for oviposition sites. *D. crucigera* breeds on twenty-one plant families, while *D. grimshawi* utilizes approximately ten.

On Kauai and Oahu, where *D. crucigera* is found together with *D. grimshawi*, the latter is a specialist, laying its eggs only on the plants of one genus. But on the islands of Maui, Molokai, and Lanai, which *D. crucigera* has not colonized, *D. grimshawi* is a generalist. This suggests that competition for food resources is responsible for the niche separation between these species. An evolutionary trend from generalism to specialization is in keeping with the popular ecological thought that populations will evolve toward the optimal use of the available resources. Geologic and behavioral data, however, indicate that the Kauai *D. grimshawi* population is the most ancestral and the one from which the generalist populations on Maui, Molokai, and Lanai evolved. The implication is that, in this case, evolution has occurred in the reverse direction.

In the laboratory, the generalist females will oviposit in vials containing our standard laboratory food medium, but females of specialist populations will only lay eggs on or adjacent to a piece of properly rotted bark of their specific host plant. This behavioral characteristic allows us to distinguish between generalist and specialist females. Thus, we are able to investigate the genetic basis of ovipositional behavior by testing hybrid females for host specificity. Furthermore, hybrid females resulting from generalist and specialist crosses are nearly all fertile, and the males are at least partly so, making it possible for us to estimate the number of genes governing this behavior.

We have only recently begun to investigate the selective forces and the genetic elements involved in the evolution of host specificity. However, some preliminary data have been obtained from hybridizations carried out on several populations of *D. grimshawi*. These data indicate that the genetic basis of the generalist ovipositional behavior is dominant to that of the specialist, and that this behavior may be regulated by a single gene. If such important behavioral components are regulated by simple genetic elements, adaptive shifts can occur rapidly and provide a beginning for the speciation process.

In summary, it appears that the Hawaiian Drosophilidae have the genetic capability to differentiate and adapt rapidly to the diversity of environmental conditions present on the islands of Hawaii. Although chance events may be primarily responsible for rapid behavioral changes during the early stages of colonizations, selective forces are probably largely responsible for directing change when the population size becomes large and competitors and predators are encountered more frequently. Indeed, our studies indicate that a small number of genetic elements may be responsible for changes in both ovipositional behavior and courtship behavior and that such changes may be partly responsible for the remarkable speciation among these flies. An exceptional capacity for rapid genetic change helped the Hawaiian Drosophilidae successfully colonize the islands of Hawaii.

8

A Trilobite Odyssey

NILES ELDREDGE
and
MICHELLE J. ELDREDGE

During the last 600 million years, ancient seas have flooded North America time and again. The hardened sediments of those ancient seabeds compose much of the underpinnings of our modern landscape. Along highways and railroad cuts, and in streams and quarries all over our continent, exposed rocks provide clues to the distribution and nature of these old seas. Living things, too, left stony traces entombed in the sediments they once swam over, walked on, or burrowed into.

These fossils show evolutionary change in two ways. They change through time in the vertical sequence of layered sedimentary rocks. Also, each species of fossil animal shows variation from place to place in the different environmental settings it encountered as it spread throughout the sea and lived on through millions of years. Evolutionary change is nothing more than variation in both time and space.

Traditionally, paleontologists have emphasized time over geographic distribution as the more important element of the evolutionary process. Strong evidence suggests, however, that evolution is not simply slow, steady change of an entire species through long periods of time. Rather, our work with several extinct invertebrates has shown that a species can exist relatively unchanged for millions of years. Real evolutionary change takes place when populations of a species become geographically isolated, allowing them to evolve in different directions from the parent species. Many of the anatomical differences between a new daughter species and its parent were established prior to isolation when the local population adapted to its specific environment.

The fossil record is full of apparently sudden evolutionary jumps, where a parent species is followed by its daughter species without intermediate fossil links connecting the two. The traditional explanation for such jumps is an incomplete fossil record, but our findings contradict tradition. When conditions permitted, animals that had evolved far away and thousands, if not millions, of years previously, migrated to territories formerly occupied by their ancestors. The sudden jump effect in any one locality actually reflects the sudden appearance of a migrant that had already evolved elsewhere. Thus, evolution requires space as well as time, and variation through space governs variation through time.

The seas of the Middle Devonian period, which waxed and waned for nearly 10 million years some 400 million years ago, provide an excellent backdrop for evolutionary studies. Located near the Devonian equator, these warm, shallow seas harbored a vast array of aquatic life, and outcrops preserving remnants of the varied marine environments extend from New York as far west as Iowa. The eastern shore of these inland seas stretched from New York through Virginia, along a rugged young mountain chain whose swift-flowing rivers dumped huge deposits of sand and silt into the sea. This Devonian nearshore environment must have been much like today's familiar sandy shorelines. As the sea extended westward from shore, heavier particles of sand gradually dropped out and graded into increasingly finer muds mixed with lime particles. The farther from shore, the more limy the sediments became. The thick accumulations of limestone where the sea covered what is now the continental interior were formed mostly of the millions of shells belonging to dead organisms. Thus, traveling from east to west across the Devonian sea, one can locate himself by the nature of the sediments. Sandstone predominates near the eastern shore, grading into shale, or lithified mud, farther out. Still farther west, more and more lime appears until, in what was the center of the sea, the rocks are predominantly limestone.

The trilobite *Phacops rana,* one of the 250 or more species of invertebrate animals that lived in this sea, had a particularly interesting evolutionary history. Because trilobites became extinct some 200 million years ago, we can only guess about their mode of life and study their modern relatives such as crabs and lobsters. Trilobites, however, looked more like larger versions of their distant land relatives, pill bugs. They were usually only about one to four inches long, and like pill bugs, they had a head with antennae and eyes, followed by a series of flexible segments that ended in a tail. To feed, *P. rana* probably crawled over the sea bottom searching for bits of decayed organic matter; it swept these up with its many legs and propelled the bits into its mouth. If threatened, most trilobites could roll up into a ball like a pill bug.

When a trilobite shed its armor, the head portion was cast aside and the animal crawled out from under the remaining shell. Then, like a recently molted softshell crab, it presumably hid from predators until its new shell hardened. Most trilobites we collect are really cast-off molts, not the remains of a dead organism.

P. rana had a very large compound eye on each side of the swollen, rounded middle region of the head, prompting the name *rana,* which means frog. We paid particular attention to the eyes because they show more evolutionary change than any other part of the body. Each eye is covered with many lenses arranged in vertical rows around the eye. Twice in the long evolutionary history of this trilobite, the number of vertical rows of lenses was reduced, giving rise to a new variety of *P. rana* each time.

Starting with eighteen rows in the most primitive variety, after ten million years and two jumps, *Phacops rana* ended up with only fifteen rows of lenses in the eye. Our field work clearly showed that these two important evolutionary changes took place in the nearshore environments of the eastern Devonian sea. Both times the sea had shrunk away from the continental interior, extinguishing the more primitive variety of *P. rana.* Each time the sea encroached again on what is now the midwestern United States, it brought with it a more advanced form of *P. rana,* which had been biding its time in the east.

Our trilobite odyssey began near the eastern shore of the Devonian sea. A quarry just north of Morrisville in central New York exposes the earliest occurrence of the typical Middle Devonian fauna. We found few fossils on the quarry floor, but as we climbed up through time to higher levels, more and more fossils began to appear. On the surfaces of these higher levels, brachiopods, clams, and snails were common, preserving the delicate features of their original shells.

Phacops rana is extremely rare here, its discovery being strictly a matter of serendipity, but in the few specimens we did find, we were able to see that the population was quite variable. These oldest *P. rana* occurring in the eastern sea usually developed eighteen rows of lenses in the eye, the primitive condition for this species. We found some specimens with an incomplete eighteenth row, however, while a few others showed the more advanced condition of only seventeen rows. Here, confined to a rather small area in the eastern Devonian sea, this varied population reveals *P. rana* in transition. We concluded that the advanced seventeen-row variety evolved in the area during this brief moment of geologic time, still very early in the history of the species.

Above this fauna were nearly 1400 feet of Middle Devonian rocks, representing a span of about eight million years. Only a few miles from Morrisville, but a million years later in time, we visited a quarry on Johnny Cake Hill, once a part of the eastern sea where fossils now lie scattered richly among the trash and spent shotgun shells that litter the quarry floor. Here, all *Phacops* specimens consistently had seventeen rows of lenses. Combing the countryside along back roads, usually unpaved and unnoted on road maps, we found other quarries of even younger rocks in the area and pieced together a picture of *P. rana* essentially unchanged with seventeen rows of lenses for five to six million years.

This lack of variation in *Phacops rana* throughout such a long period of time is a remarkably graphic illustration of the tendency of organisms to persist unchanged given the persistence of their habitats. Such stasis, of course, challenges the traditional view of evolution as gradual progressive change. Only in the very youngest rocks of this area did we again encounter variation in the eyes. Near the end of its history, *P. rana* evolved an eye variety with only fifteen rows of lenses. Again we found a transitional population; this time a mixture of specimens, with seventeen, sixteen, or fifteen eye rows, briefly coexisted. Once more, evolutionary change seems to have taken place in a short period of time in the nearshore environment of the eastern sea.

We pursued our odyssey over the Devonian sea into quieter, deeper waters farther west. *Phacops* gradually became more abundant and seemed at home here in the limy offshore muds.

The area south of Buffalo, New York, is justly famous among trilobite lovers, and we found it prime collecting territory. The best exposures are natural ones in stream beds, waterfalls, and on the shores of Lake Erie.

Where Eighteen-Mile Creek meets Lake Erie, imposing cliffs rise straight up from the shore. Impressive as these cliffs are, the rocks underfoot are the real mother lode. A thin layer of ancient sediments crops out on the shore, strewn about with the modern sand and flotsam of Lake Erie. These layers are famous as Grabau's Trilobite Beds, named for the Columbia University paleontologist who studied them early in this century. And they are well named. *Phacops rana* is so common and easy to spot that hundreds of heads, tails, and even complete specimens can be gathered in half an hour. The dark hue of *Phacops* skeletons makes them conspicuous against the light background of gray shale. Here, any fossil is likely to be a trilobite since it exists almost to the exclusion of other species.

Considering the all-important eye, it was easy to determine with such an excellent sample that in Grabau's Beds, *Phacops rana* uniformly had seventeen rows. Now it remained to find out what *Phacops* was doing in the older and younger rocks of the area. As we walked along the shore, our elevation remained unchanged while the cliffs imperceptibly dipped to the south, so we found the upper beds of the cliff cropping out at shore level after a couple of miles. Thus we were able to collect *Phacops* from all the overlying rock units without resorting to Alpine gymnastics.

In the end, every rock unit sampled at Buffalo had the same story to tell: again we found *Phacops rana* with seventeen rows of lenses persisting unchanged for millions of years, so here too, a remarkable picture of stability emerges, over space as well as time, just as in the central part of New York.

Again we headed west, farther out into the Middle Devonian sea. Along the entire route between Buffalo and the region of Arkona in southwestern Ontario, the Middle Devonian rocks are deeply buried. They finally crop out in the bluffs along the Ausable River in Ontario, but only about the middle third of Middle Devonian time is represented. Here, for the first time on the trip, we saw significant geographic variation across the bottom of the sea. The Arkona specimens all had eighteen rows of lenses while those of the same age in New York had only seventeen. Apparently the more primitive variety of *Phacops rana* was able to hang on for a longer period of time in the interior sea than on the margin.

Along the bluffs at Hungry Hollow on the Ausable River, rapid erosion is tearing down the soft, calcareous shales, creating slippery tracks of light gray mud. After being cemented nearly 400 million years in the Devonian seabed, clay particles rest here momentarily in their journey back to the modern sea. Yet even at this moment, the processes of fossilization go on. At our feet, amid loose Devonian fossils, lay the exquisite remains of mayflies stuck in the dried mud. But their temporary entrapment was to be as ephemeral in the reckoning of geologic time as were their lives by the daily reckoning of man.

Collecting our way up the shaly bluff, we kept finding the eighteen-row variety of *Phacops rana* until we reached a thick bed of limestone. Here, where the rocks changed abruptly, so did the trilobites. Suddenly, after two million years without change, the more primitive variety was supplanted by the more advanced seventeen-row trilobite already familiar to us from New York. Our closest scrutiny of the shale-limestone boundary turned up no specimens intermediate between the two varieties. Rather than supposing a sudden evolutionary jump between the two varieties, we concluded that the seventeen-row *P. rana*, already living in the east, simply migrated west as time passed. Thus the seventeen-row variety arrived in the seas of the continental interior some one or two million years after it had evolved closer to shore.

The next outcrops were in Michigan along the northern shore of Lake Huron, our westernmost stop in the Devonian sea. The shales and limestones seen on the lakeshore, in riverbanks, and particularly in large commercial quarries preserve a fairly complete record of Middle Devonian time. The quarries are particularly impressive; most of them, such as the quarry operation just north of Alpena, blast out huge amounts of limestone for cement making. We had mixed emotions whenever we saw a large dump truck hauling tons of fossiliferous limestone off to the crusher. On the one hand, we witnessed the daily destruction of thousands of fossils; on the other, without the quarrying, exposures of fossiliferous beds would be fewer, possibly even nonexistent, and we would end up with a much poorer picture of Middle Devonian life in Michigan.

From time to time in the past, what is now the southern peninsula of Michigan seems to have been cut off from complete communication with the seas elsewhere in the continental interior. As a consequence of this isolation, the organisms living within the Michigan Basin tended to diverge slightly from their relatives elsewhere. Here again, geographic isolation played an important role in evolutionary diversification.

The Michigan Basin's peculiar environmental and geographic history directly affected *Phacops rana*. Another trilobite, *P. iowensis*, which is not a close relative, dominated the Michigan seas so completely that *P. rana* appears only sporadically and sparsely. *P. iowensis* seems to have lived mainly in the Michigan Basin throughout most of its existence, venturing outside of the area only at odd intervals. The Michigan Basin was its turf, and *P. rana* a rival interloper. The different *P. rana* seen in Michigan, then, reflect occasional infiltration, rather than continual habitation there.

All *Phacops rana* specimens that we found in the lower half of Michigan's Middle Devonian sequence had eighteen rows of lenses in the eye. They were replaced by the seventeen-row variety, without any sign of gradual modification, in a thin unit corresponding in age to the thick limestone we found in Ontario. Thus far, the story of *P. rana* in both Ontario and Michigan was identical: persistence of the primitive variety over a span of time, with an abrupt and simultaneous switch to the seventeen-row type.

All but the very latest portion of the more recent half of Middle Devonian time is missing in Michigan. Rocks of latest Middle Devonian age crop out along the shores of both Lake Huron and Lake Michigan to the west. At Lake Huron, only *Phacops iowensis* is present, but along the shores of Lake Michigan a variety of *P. rana* with fifteen eye rows is present. These specimens are very similar, and presumably closely related, to those from the uppermost part of the sequence in eastern and central New York. It appears that again a wave of migration of an advanced eye variety invaded the seas of the continental interior from the east.

Although we sampled many other formations over the interior of the United States and along the southern Applachians, the localities we have touched on here provided the key for understanding the evolutionary pattern of *Phacops rana* during its ten million years of existence. Many changes occurred in *P. rana* throughout its spatial and temporal distributions, but the changes in number of rows of lenses ultimately proved to be the most significant. The pattern was so clear that, given the age and location of a rock, we could predict which eye variety of *P. rana* might be present. Conversely, given a specimen of *P. rana*, we could position it in time and space simply by examining the eyes.

A very simple picture emerged. Twice the great continental sea dried up and extinguished the *Phacops* living in it. But twice a more advanced type of *Phacops* with fewer eye rows had evolved on the eastern margin of the sea. Each time the continent was flooded, a different variety of *Phacops* spread west with the advancing sea.

All over the continental interior, the most primitive, eighteen-row variety of *Phacops* was present in the older half of the Middle Devonian. Meanwhile, the more advanced seventeen-row variety of *P. Rana* was thriving in the east. We know the sea shrank away from the interior near the end of the first half of Middle Devonian times; this is clearly implied by the absence of sediments of this age anywhere in the Midwest. With the disappearance of their habitat, the primitive eighteen-row variant of *P. rana* simply vanished.

The Centerfield Sea, named for the Centerfield Formation of New York, reflects the first widespread inundation of the continent after this major withdrawal. This sea left traces in New York, the southern Appalachians, Ontario, Ohio, Michigan, and Indiana. Its sediments are easily identified by the distinctive fauna of corals and brachiopods, and everywhere that the remnants of this sea yield *Phacops rana,* only the seventeen-row variety is present. The conclusion must be that the advanced form is a migrant from the east.

This migrant, however, did not simply evolve after the extinction of its primitive relative to the west. On the contrary, we found it in very old rocks in New York and the Appalachians. For approximately two million years, the seventeen-row variety had lived in the east, unable to expand westward over the continental interior because of the continued existence of the eighteen-variety already out there. Only after this primitive type became extinct could its relative gain a foothold in the western seas.

A similar episode—when the seventeen-row variety gave rise to the fifteen-row variety, again in the east—occurred near the end of the history of *Phacops rana*. This time, another, even longer withdrawal of the continental sea preceded the repopulation of the interior by a more advanced eye variant. Fifteen-row *P. rana* are known from the very latest Middle Devonian rocks of New York,

Michigan, Wisconsin, and Iowa. Their story was short, however, for soon the sea disappeared once more, and by the time the continent was again flooded, *P. rana* had become entirely extinct.

Perhaps the most amazing feature of the entire *Phacops rana* story is its stasis—a persistence against change—through vast amounts of time. Contrary to popular belief, evolutionary change seems to occur infrequently, and usually in small, isolated populations in a short span of time. The bulk of a species' history is stasis, and there is no inexorable, progressive evolutionary march through time.

PART 3
ECOLOGY, BEHAVIOR, AND EVOLUTION

Ecological and behavioral themes run deep in the literature on evolution, and have been prominently featured in the pages of *Natural History* for years. Yet ecologists spend most of their time trying to analyze how organisms interact with each other and the physical aspects of their environment. For such functional studies of living systems, evolutionary concepts often play a negligible role. Similarly, animal behavior is an object of study in and of itself.

Yet, equally obviously, the ecological arena, where the game of life is played out on a day-to-day basis, is where all those adaptations are actually *used*. In a very real sense, evolution is about the development, maintenance, and modification of ways of making a living, of playing the economic game of life and continuing to pass that information along via reproduction. Viewed this way, natural selection is the interface of ecology and the genetically based rules for living: degree of success at the economic game determines, more or less, how successful an organism will be when it comes to reproductive matters.

And behavior, too, must evolve. Behaviors are as much a part of an organism's phenotype as are its skeleton and biochemistry. Those behaviors that are strongly genetically based are perhaps more easily seen as adaptations in a classic, Darwinian sense. But learned behaviors, with but a vague basis in genetic information, certainly also evolve and function as adaptations.

The essays in part 3 seek to forge definite links between evolutionary theory and ecology and the study of behavior. In the first essay, John A. Wiens gives a lucid exposition of a rather controversial and startling thesis: the evidence for competition between local populations of different species is not as compelling as we have all, for over a century, assumed it should be. Wiens briefly alludes to the mathematically centered work in theoretical ecology that pinpointed the effects of competition precisely enough to enable field ecologists to formulate predictions and go out to nature to test the theories. Lo and behold—according to Wiens and a substantial number of his colleagues—communities often appear *not* to be structured as if the presence or absence of a species affects the population size of other species. And different species often do take the same sorts, and sizes, of food items—again, contrary to the old axiom that different species coexist because each is specializing on a different food, with little or no overlap in dietary preferences. Nature, according to Wiens, is not so rigidly, tightly structured as we have been led to expect. According to traditional evolutionary biology, competition is one of the well springs of natural selection; here the message of the field ecologist is a serious challenge to conventional evolutionary wisdom.

Richard D. Alexander, who has become a noted proponent of the relatively young field of sociobiology, next provides us with an analysis of the evolution of cricket chirping. Focusing on the family Gryllidae, Alexander is concerned to integrate a complex behavioral feature—chirping—with what is known of the anatomy, physiology, evolutionary relationships, and fossil record of crickets.

Cricket chirps are species-specific: with rare exceptions, no two cricket species chirp alike. *Within* species, however, there is no deviation (other than variation induced by temperature changes). The moral: cricket chirping is an adaptation that fosters successful matings between members of the same species. Alexander's analysis fits in beautifully with the theme of the previous chapter on speciation: if species are reproductive communities, each species must have some unique feature in its reproductive adaptations (what biologist H. E. H. Paterson has called the "Specific Mate Recognition System"). Speciation, minimally, must involve at least some change in this system—in the case of crickets, chirping behavior.

With Collias' essay on the evolution of nest building in birds, we return to the theme of adaptation. His concern is primarily functional: why do nests look the way they do? What is the adaptive significance of all those various shapes, sizes, and degrees of complexity? Surveying a broad spectrum of bird nest building, this article on behavior would fit as well in part 1 on adaptation. Notice, too, how Collias (like Kaneshiro and Ohta in the middle essay of part 2) relies on the notion of interspecific competition in building his argument—once again, in good Darwinian tradition, but counter to the views expressed by Wiens!

9

Competition or Peaceful Coexistence?

JOHN A. WIENS

Examples of competition between species seem to abound in nature. A calliope hummingbird feeding at a nectar-rich flower in a mountain meadow is apt to be supplanted by a larger broad-tailed hummingbird, which, in turn, may be chased from the flower by a rufous hummingbird. Deer mice on islands that have no meadow mice seem to occupy a broader range of habitats and exploit more types of foods than do deer mice in adjacent mainland areas where meadow mice are present. As one travels up a mountainside in western North America, the scrub jays common at lower elevations may suddenly disappear, replaced by ecologically similar Steller's jays.

These kinds of observations have led many ecologists to conclude that competition between species is commonplace and that it determines, to a great extent, how natural communities are put together. According to this view, species are likely to compete if they require similar resources, such as food, habitat, or breeding sites, and if those resources are in limited supply. This competition will lead to the exclusion of one species by another or, over evolutionary time, to divergence in the species' use of resources until competition is minimized. Competition thus limits the number and kinds of species that may coexist in an environment.

These views are by no means new. To Charles Darwin, competition between and within species was a fundamental component of the "struggle for existence." In *On the Origin of Species,* he noted: "We have reason to believe that species in a state of nature are limited in their ranges by the competition of other organic beings quite as much as, or more than, by adaptation to particular climates." Elsewhere in this work, Darwin provided examples of such competitive exclusion:

The struggle will generally be more severe between species of the same genus, when they come into competition with each other, than between species of distinct genera. We see this in the recent extension over parts of the United States of one species of swallow having caused the decrease of another species. The recent increase of the missel-thrush in parts of Scotland has caused the decrease of the song-thrush. How frequently we hear of one species of rat taking the place of another species under the most different climates! In Russia the small Asiatic cockroach has everywhere driven before it its congener. One species of charlock will supplant another, and so in other cases. We can dimly see why this competition should be most severe between allied forms, but probably in no one case could we precisely say why one species has been victorious over another in the great battle of life.

In the early 1960s, interest in the role of competition was stimulated by the development of mathematical theories of the ecology of communities. Numerical models attempted to describe the degree of difference needed between species to permit their coexistence, how resources might be subdivided between species, how the configuration of a habitat might affect the number of species present, how a species might expand its use of resources in the absence of a competitor, and so on. In part to make the mathematics tractable, such models contained the assumption that the communities they described were in equilibrium—that is, that the species making up a community were in balance with their resources and with one another. This theoretical work generated interesting questions and neat predictions about competition and communities and sent ecologists scurrying into the field and the laboratory to test the predictions and answer the questions.

Over the past decade, my student John Rotenberry (now on the faculty at Bowling Green State University) and I have explored these ideas. We have concentrated on communities of breeding birds at a number of locations in western North America, scattered from the prairies of the Great Plains to the shrub-grass mixtures (shrubsteppe) of the Great Basin. We chose these environments for several reasons. First, they are open. Prairie grasses and shrubsteppe plants, such as sagebrush, rarely exceed knee height; shortgrass prairies often resemble a well-trimmed lawn. We can easily see what the birds are doing. Second, these environments generally support relatively few species of grasses or shrubs and their physical structure is simpler than that of a multilayered forest. Consequently, different species of birds should be compelled to compete for the same resources more often than in more complex habitats. Third, the production of plant material and insects is relatively low, which we believed was another reason the birds might often have to compete for food. Finally, the habitats support few breeding bird species (generally two to eight per twenty-five acres), and the possible pathways of interactions among the species should thus be much less complex and easier to study than in habitats with many species.

We fully expected either to find competition going on in these communities or to be able to document patterns confirming that competition had existed in the past. As the research progressed, however, these expectations proved to be naive. We now think that direct, ongoing competition is infrequent in these systems and that it may have relatively little to do with the organization of the bird communities. But I am getting ahead of my story.

The communities of breeding songbirds in the grasslands and shrubsteppe of North America are dominated by a few characteristic species—eastern meadowlarks, dickcissels, and grasshopper sparrows in the lusher eastern grasslands; western meadowlarks, horned larks, and longspurs in the shortgrass prairies; and sage thrashers, sage sparrows, and Brewer's sparrows in the shrubsteppe. These and the other songbirds that are present are ecologically similar. All of them forage on the ground or in low vegetation and feed primarily upon insects and other arthropods during the breeding season. Most nest directly on the ground, in small flowering plants, or in shrubs, and males proclaim their territorial holdings by singing from exposed, elevated perches or in flight. Some, such as the meadowlarks and the dickcissels, mate polygamously, but most form monogamous pair bonds that are maintained through the breeding season. In the shrubsteppe of Oregon, where we have watched the birds closely for several years, adults seem to return to previous breeding locations for several successive years, but young birds apparently settle elsewhere to breed. This pattern is likely to hold in other shrubsteppe and grassland habitats as well.

Competition theory suggests that, such general similarities notwithstanding, these coexisting species must be adapted to grassland and shrubsteppe conditions in different ways. One way to accomplish this necessary separation among the species is through differences in diet. If, as is generally assumed, populations are normally limited by their food supplies, then the feeding habits of coexisting species should reveal something about their relationships, including possible competitive interactions. To test this prediction, we sampled the diets of breeding species at four widely spaced locations for two, and in some cases, three, successive years. Our sampling method involved collecting the birds and identifying food fragments in their stomachs. Unexpectedly, the patterns that emerged seemed muddled. Some of the coexisting species did differ substantially in the types of prey they consumed but often not consistently from year to year. At a South Dakota prairie, for example, both 17-gram grasshopper sparrows and 32-gram horned larks gleaned large numbers of moth and butterfly larvae from the vegetation one year. The following year, however, the larks switched to a diet composed largely of chenopod seeds, while grasshopper sparrows preyed almost entirely upon grasshoppers. Moreover, in some years the diets of grasshopper sparrows overlapped extensively with those of the substantially larger (110 grams) meadowlarks.

Neither insect taxonomy nor avian body size thus seems to be a good measure of differences in diet among members of these communities. Because birds use their bills to obtain food, however, bill size might be more closely related to diet. Community theory, in fact, predicts that species should be evenly spaced along a bill-size gradient, each successively larger species having a bill about a third again as large as that of the next smaller species. Species with bill-size differences much smaller than this 1.3:1 ratio presumably will have extensively overlapping diets and will thus compete, leading to the elimination of one or the other. On the other hand, if two adjacent species on the gradient differ by a substantially greater amount, a third species, with an intermediate bill size, should be able to invade the community. The community should thus rapidly attain this even-spacing configuration, but the bird communities in grasslands and shrubsteppe do not. In these habitats, the spacing of species along a bill-size gradient is outrageously uneven, and there are substantial gaps in the sequence. In the South Dakota prairie, for example, grasshopper sparrows had a bill only 1.03 times longer than that of chestnut-collared longspurs. The bill of horned larks was 1.28 times the size of that of grasshopper sparrows (reasonably close to the predicted ratio), but the bill of the next and largest species in the sequence, the western meadowlark, was 2.26 times the length of a lark's bill. Other communities we looked at also seemed less consistently patterned and less fully packed with species than the theory led us to expect. Like body size, bill size failed to turn up clear evidence of differences caused by competition.

We also investigated the size of the food items eaten by the birds. Although the kinds of insects or seeds eaten by different species overlapped, perhaps the sizes might not. Coexisting species could then diverge ecologically by specializing on prey of different sizes irrespective of taxon. One of our study sites, a west Texas shortgrass prairie, seemed to support this possibility. There, the breeding species (horned larks, grasshopper sparrows, and western meadowlarks) differed substantially in the sizes of the foods they consumed. Furthermore, they differed in strict accordance with their body- and bill-size rankings. Other locations, however, produced conflicting results. At our South Dakota location, the diets of all species included prey of various sizes and overlapped extensively. The birds in the southeastern Washington shrubsteppe ate food items of virtually identical sizes. At both sites, there were considerable differences in bill and body sizes among the birds.

Excited by these findings and wanting to learn what, if not bill or body size, was determining the birds' diets, Rotenberry subjected the Washington shrub-steppe birds to greater scrutiny. By sampling their food habits frequently, he demonstrated that each species changed its feeding habits dramatically through the breeding season and that the diets of the different species changed in tandem. For example, as the sage sparrows switched from a diet dominated by beetle larvae, grass seeds, weevils, and grasshoppers in April to one dominated by weevils and lepidopteran larvae in May, so did the larger horned larks. Overall, the species seemed to be responding opportunistically and similarly to seasonal changes in the availability of different types and sizes of food. This certainly is not the sort of pattern one would expect of a set of species locked in intense competition over food.

Important as food is, it is only one of the resources over which competition might occur. Habitat—a place to live—is equally vital. If the habitats of coexisting species differ consistently, even in subtle features, this might circumvent competition even among species with similar diets. After all, if the species obtain their food in different places, what does it matter if the prey are of the same types or sizes?

To determine the relations of birds to their habitats, ornithologists generally look at habitat in structural terms, measuring such features as vegetation height and density, understory coverage and stratification, grass cover, and litter accumulation. The low vegetation in shrubsteppe and grasslands make such features easy to measure, and we recorded habitat structure in many locations. This information allowed us to ask, Are there clearly defined sets of co-occurring species that are closely related to features of habitat structure? If so, they might represent the groups of competitively adjusted species that theory leads us to expect.

The answer to the question turns out to be yes or no, depending on scale. When the entire spectrum of environments from midwestern tallgrass prairies to northwestern shrubsteppe is considered, well-defined groupings of co-occurring species do exist and are clearly related to large-scale variations in habitat structure. The distribution and abundance of dickcissels and grasshopper sparrows, for example, are tightly linked to tall vegetation, extensive grass cover, and a well-developed litter layer—features of tallgrass prairies. Sage sparrows and sage thrashers, on the other hand, avoid grassy areas and instead occupy habitats dominated by shrubs, where vegetation is patchily distributed and large areas of bare ground are common—typical shrubsteppe conditions.

On this large a scale, our findings offer little more than intuitive association of birds with certain habitats, which any practiced bird watcher can make. To provide significant evidence that these sets of species have been organized by competitive interactions, we would need to discover the same bird-habitat relations when the focus is restricted to small-scale variations within one habitat type, such as the shrubsteppe. When we looked at a variety of shrubsteppe locations, however, we detected no consistent sets of co-occurring species. The species, instead, varied in abundance and were distributed independently of one another. Moreover, at this scale we found that in most cases the associations between individual species and features of habitat structure were weak at best. Several methods of analysis indicated that, overall, less than 17 percent of the variation in the distribution and abundance of the birds in these areas could be explained by habitat structure.

Clearer patterns of bird-habitat association emerged when we considered the species composition of the vegetation, instead of its structural configuration. Within the shrubsteppe, the abundance of sage sparrows clearly varied in accordance with the amount of big sagebrush present, while sage thrashers and Brewer's sparrows varied independently of sagebrush but seemingly avoided areas where small spiny shrubs, such as hopsage or budsage, were common. These associations were not strong, but they do indicate that the plant species composition of an area may be more important to the birds than has been thought. Desert shrubs vary in the chemical composition of their leaf tissues, and these chemical differences may influence the abundance and variety of insects present on different plant species. If this is so, some apparent associations of plants and insect-eating birds may be reflections of the birds' food preferences.

In any case, the birds that breed in America's shrubsteppe and grasslands exhibit little regard for the predictions of ecological theory. Variations in the population size of one species in an area are largely independent both of the presence or absence of other species and of variation in habitat features. Coexisting species appear to use resources more or less opportunistically. We find little evidence that they are currently much concerned about competition with one another or that competition in the past has led to an orderly community structure.

When observations of nature do not match the predictions of a theory, we must ask why. In this case, one possibility is that the theory is inappropriate for grassland and shrubsteppe bird communities. A basic assumption of competition theory is that the communities one observes in nature, like the mathematical community models one can create, are in equilibrium. Is this assumption violated in our systems?

As any farmer or rancher will tell you, both midwestern grasslands and western shrubsteppe are extremely variable and unpredictable environments. From one year to another, the weather can vary from deluge to drought, and the effects on natural vegetation, as well as on croplands and pastures, can be profound. The weather in these environments is certainly not in equilibrium, but are the bird communities? The answer hinges in part on how long it takes the birds to respond to changes in their environment. If they take much time to adjust, then the interactions between species may be altered and the effect of competition on the community will be diffused. If, however, the birds respond with no appreciable time lag, they might maintain a rough balance, or equilibrium, with resource levels. In that case, competition could still have a strong influence on community structure.

Our shrubsteppe studies included one of the driest years on record in the region, followed by two abnormally wet years. This provided an opportunity to observe how the vegetation and the birds reacted to such dramatic fluctuations. The response of the vegetation at our sites was clear-cut: in the two wet years that followed the drought, annual plants and grasses flourished. The overall height of the vegetation and the extent of ground cover increased; the amount of bare ground and patchiness decreased. The bird populations also varied, but none of the species changed in a way that was clearly associated with changes in habitat structure.

Another alteration in habitat, this time a result of human activities rather than weather, provided some perspective on why the birds may not closely track such variations. At one of our sites in southeastern Oregon, state and federal agencies applied herbicides as part of a "range improvement" program. The following fall, the native shrub, sagebrush, was disked and an exotic bunch grass, crested wheatgrass, planted. Because we had monitored this site for three years before the application, we could record the response. The vegetation, of course, was decimated—sagebrush coverage decreased from more than 25 percent of the ground area to less than 2 percent, and no vegetation taller than eight inches remained. Despite this, sage sparrows, which had in past years shown a clear preference for sagebrush, returned to the site in about the same numbers as before the treatment. We think that these birds had previously bred on the site and that the urge to breed in a traditional location overrode the tendency to select an appropriate habitat. Our continuing studies should record a decline in sage sparrows as these adults die or give up and move elsewhere. This example indicates that time lags in the responses of individuals to environmental changes can complicate attempts to compare natural systems to ecological theory.

In variable environments, such as grassland and shrubsteppe, populations may often be out of phase with their resources. This may provide the setting for feast or famine situations: periods of benign environmental conditions, when resource supplies may far exceed demands, may be punctuated by periods of sharply reduced resource availability. During these ecological crunches, the supplies of resources may be so limited that competition among the species intensifies, leading to precisely the sorts of consequences predicted by theory. During the intercrunch intervals, however, the relative superabundance of resources may render competition unnecessary.

All of this suggests that competition is not the ubiquitous force that many ecologists have believed. Certainly, it does occur in some situations. Careful studies of groups of hummingbirds feeding on nectar, for example, have clearly documented competition between the birds, as well as between the birds and bees. Competition may be more likely to exist and easier to perceive in stable environments. In unstable environments, however, population sizes may be unrelated to immediate resource conditions, and assemblages of species may often not express the relationships that theory says they should.

Upon reflection, these statements seem to make good sense. Why, then, have ecologists since the time of Darwin been so preoccupied with competition, and why, in thinking about competition, have so many assumed that nature is more or less in equilibrium? Part of the answer is that we have used simplified theories in an attempt to gain some understanding of nature. But our views have also been influenced by their cultural context. Science does not develop in a vacuum but is a mixed product, influenced by previous findings and ideas in the discipline and by the prevailing world views of the society in which it grows and matures. The notion of equilibrium is deeply embedded in Western culture. It derives from Greek metaphysics, which portrayed the universe as ultimately ordered and balanced, and it is expressed in the commonly accepted notion of a "balance of nature." Competition also occupies a central position in Western culture—witness its expression in sports, economics, space exploration, international politics, or warfare. Little wonder, then, that community ecologists expected that the species they studied would be in balance with one another and with their resources, and that the primary factor organizing communities would be competition. After all, we have grown up immersed in such a world view. But now, the birds of grasslands and shrubsteppe seem to be telling us that nature may not always be this way. Darwin's "great battle of life" may be fought in skirmishes that are interspersed with periods of relative peace.

10

The Evolution of Cricket Chirps

RICHARD D. ALEXANDER

Crickets must have been a particular source of curiosity for as long as there have been humans to be curious. Poets, philosophers, scientists, and primitive peoples—all have left some special indications of their interest in one kind of cricket or another. In addition to verses about crickets, written in many languages, one could cite the antiquity of cricket fighting as a sport in the Orient; the almost universal fame of Milton's "cricket on the hearth"; and the practice, developed independently in several parts of the world, of keeping crickets for food, for their songs, and for driving away "evil spirits." Even today, with so many millions of the world's population concentrated inside cities of macadam, steel, and concrete, the producer of a television show or motion picture can make his scenes nocturnal with only a token drop in light intensity if he simultaneously adds the chirp of a cricket in the background.

The source of the cricket's charm is obvious. To eliminate it we would need only to silence him, to take the acoustical dimension out of his life. But this would not be a simple exorcism. As the jigsaw puzzle of cricket life slowly assumes shape through a continuing series of investigations on behavior, classification, structure, and physiology, it is increasingly evident that a certain well-known American biologist could not have been more wrong when he wrote that, like the clanking of a knight's armor, cricket chirps were little more than the frictional creakings of an animal with an external skeleton. Indeed, those cricket groups that have lost their ability to stridulate, and along with it their hearing organs, have changed their lives and their body forms so drastically that they are excluded by some insect taxonomists from the elite body of "true" crickets.

Fossil evidence indicates that the crickets (family Gryllidae) became a separate evolutionary line some 150 to 200 million years ago, probably during the Jurassic Period, coincident with the heyday of the dinosaurs. Their acoustical system is even older than that, since their nearest relatives, the katydids and long-horned grasshoppers (family Tettigoniidae), have the same tympanal auditory organs on their front legs and the same stridulatory device on their front wings. It is possible that this is the oldest acoustical communicative system still in existence, and certainly crickets and their relatives were among the first animals to be heavily involved in transforming the previously silent terrestrial environment into the bedlam of noise it had become long before the first humans appeared on the scene.

Crickets are the master musicians of the insect world. They have, within a single species, at least as many different kinds of acoustical signals as any other kind of insect; they produce some of the loudest of all animal signals (over 100 decibels at distances of a few inches); and they are the only animals known to be capable of producing a "pure" frequency with a stridulatory, or rubbing, device. (Pure frequency means that a single frequency so dominates the sound that all others are inaudible and, for practical purposes, insignificant. Such a sound is difficult to achieve. Not even an electronic audio-oscillator, for example, has been able to produce an absolutely pure frequency.)

The cricket stridulatory and auditory organs are complex devices that have evolved together as a functional unit for a long time; they could not have appeared through a single change, or even a few mutations, but had to develop through a long sequence of small, step-by-step alterations. There are no fossils of rudimentary versions of these devices, but there is evidence of their precursors among the living relatives of crickets. Nonacoustical relatives of crickets have a large clump of sensory cells, called the subgenual organ, in the same general location on the forelegs as the crickets' auditory organ. The cells apparently function as proprioceptors, supplying information to the insect's central nervous system about the position of the leg. These proprioceptive cells are believed to be the forerunners of the tibial auditory organ, and we can speculate that the device may have passed through a vibration-perceiving stage, during which it was sensitive only to transverse waves that were transmitted through the substrate. Subsequently, at one spot the cuticle thinned and special membranes appeared, making the spot gradually more sensitive to the air-transmitted, longitudinal waves that we call sound.

The origin of the stridulatory device on the male cricket's forewings can also be reconstructed, in the absence of fossil evidence, through comparison of living crickets and their relatives. Most of the modern groups of winged insects that have left the oldest fossil records mate with the female climbing on the male's back. Nearly all modern crickets still mate this way, and comparative study correlating structure and behavior suggests that all their ancestors, back to the cockroach line from which they diverged during Paleozoic times, mated in the same fashion. Most cockroaches still start copulation this way (although, like some crickets, they finish the act end to end), and like the male crickets, male cockroaches raise their forewings during courtship, exposing chemical areas on their backs that attract the female into the mating position. But cockroaches never developed prominent stridulatory devices, even though some of them rustle their wings audibly during courtship, and they never developed auditory organs on their forelegs. While crickets and katydids were becoming acoustical, the cockroaches were elaborating chemical and tactual stimuli, and so they remained cockroaches.

It seems beyond question that cricket stridulation originated from the wing lifting and vibrations of courtship. If the auditory organ also evolved in this context, and it seems probable that it did, we may wonder if, during the vibration-perceiving stage postulated above, the source of vibration might not have been the shaking and wiggling body of the male as he vibrated his lifted wings. The advantage of thus providing a vibratory stimulus to the advancing and mounting female could have resulted in a stridulatory ability that would add to the vibrations (and incidentally produce acoustical effects), even before the appearance of auditory ability. This, in turn, could have set the stage for elaboration of the auditory organs and completion of the transition to acoustical living for the ancestor of crickets and katydids.

Although this hypothetical sequence involves a great deal of circumstantial evidence and speculation, the facts fit together so beautifully that, in the absence of any evidence to the contrary, the whole idea seems quite reasonable.

But to account for the appearance of the auditory and stridulatory apparatus is only the beginning of the story. These devices have been around for 150 million years, and during that time their actual structure has changed in only relatively minor ways. Crickets and katydids diverged very early and developed quite different sorts of sounds: crickets, their clear, whistle-like notes; and katydids, their lisps and clicks that are often almost of the nature of white noise (containing an extremely wide spectrum of frequencies). The auditory and stridulatory devices of the two families are correspondingly different. But there are some 2500 known species of crickets and even more katydids. Hardly any two species have the same sounds in their repertoires. Here, in the analysis of signal diversity, are found the interesting and most difficult questions regarding evolutionary change. In many cases there are no differences at all between the auditory or stridulatory devices of species: song differences depend instead on some unknown variation in their central nervous systems or possibly in their muscles.

You may ask why species differences in cricket stridulations should always be attributed to evolutionary change. After all, humans make different sounds in, say, China and England, but a Chinese baby reared in London would speak perfect English, and a British baby reared in China would speak perfect Chinese. Not so with crickets. So far all of the environmental manipulations that entomologists have been able to dream up, short of actual mutilation, have had no effect on the kind of chirp a cricket makes, or the various chirps to which it can respond. If it does chirp, it gives the right chirp for its species, and it does so the first time it tries, after only a few raspy starts. This is really no surprise, for unlike birds and mammals, or even frogs and toads, most young crickets do not hatch from the egg until long after all individuals of the previous generation have died. This means that there can be no "culture" at all in cricket chirping. Differences among species and differences within species, in both signals and responses, have—in every case tested—been shown to be the result of genetic differences. The only exceptions are song variations such as those brought on by temperature (crickets are cold-blooded animals), and even here, the ability to respond also changes with temperature. Thus, a female cricket can recognize a singing male of her own species only if he is approximately as warm or as cold as she is.

This rigidity in the acoustical behavior of crickets typifies much of the behavior of arthropods. It is not merely a failure to evolve learning; it is another *direction* of natural selection. Selection has been minimizing the chances that the kinds of sounds a cricket makes will be influenced by sounds it hears; there are too many alien sounds in a newly adult cricket's environment, and too few chances of hearing another cricket at just the right moment. This does not make the cricket's chirp any less a product of both hereditary and environmental factors: every characteristic of an organism, after all, depends upon both factors. But the particular environmental factors involved in the development of a cricket's chirp are much more difficult to identify than some of those influencing, for example, bird songs or human vocalizations, and they are evidently less variable among the environments of different individuals of the same species.

The evolutionary elaboration of diversity in cricket signals has taken place in two contexts. On the one hand, species have begun to produce acoustical signals in new situations, and by evolving the ability to make use of such innovations, they have increased the number of effective signals in their repertoires. On the other hand, whenever speciation has occurred, the resulting species have evolved different repertoires. We have tape-recorded about 350 signals from a total of more than 200 cricket species, in 10 subfamilies and 50 genera, brought together from all parts of the world. In this entire assemblage there are only three pairs of species, one group of three species, and one group of four species that have any identical signals among them.

What kinds of variations occur among cricket signals? First, crickets can never produce sustained tones: because they make all their sounds by oscillatory motions of the forewings, their sounds must be successions of "pulses," each produced by one stroke of the forewings. So far, all crickets recorded appear to sonify only during the closing stroke of the wings (against the slope of the teeth on the stridulatory file) and, except rarely, to open the wings silently.

A cricket can produce one pulse at a time or a few together (a "chirp" in either case, by common definition), or he can deliver a long series of pulses that may be regularly, irregularly, or not all interrupted (a "trill"). The fastest pulse rates known are about 250 per second at 80 degrees Fahrenheit in some North American bush crickets in the subfamily Eneopterinae. Cricket sounds can also vary in frequency (cycles per second, roughly equivalent to "pitch" in human terms), and this usually relates to the size of the insect. Ordinary house and field crickets, about one-half inch long, all chirp at 4000 to 5000 cycles per second; some of the tiniest crickets, 1/16 to 1/8 inch long, chirp at more than 10,000 cycles per second; and the large mole crickets, an inch or more in length, chirp at about 1,500 cycles per second, which is about the pitch of the third G above middle C on the piano.

With one exception, all known cricket chirps are associated either directly or indirectly with the reproductive function. A student of mine, Daniel Otte, has found that one of the giant, burrowing *Brachytrupes* species, known as "bull crickets" in South Africa, makes an "alarm" or "disturbance" squawk, as do many other insects, when seized or harassed in its burrow. Aside from this exception, six functional kinds of cricket signals have been identified, and a single North American species, the short-tailed cricket, *Anurogryllus muticus,* appears to possess all of them. This is a greater variety of acoustical signals than is known for any other kind of insect, or for any fish, amphibian, or reptile, and even for many birds. Actually, relatively few mammals have been shown to have as many as six different acoustical signals, although this surely is because of inadequate study in practically all cases.

The six acoustical signals functional in the reproductive behavior of crickets may be described as follows:

1. *The calling song* attracts sexually responsive females from considerable

distances—outside the range of other senses—and elicits aggressive behavior in hyperaggressive males. It is almost certainly important in the spacing of territorial, singing males.

2. *The courtship song* stimulates the female to move forward and into the mating position.

3. *The aggressive sound* causes other males to fight, chirp, or retreat, depending on the situation.

4. *The courtship interruption sound* has no proven function. It may call females back to males after accidental separation.

5. *The postcopulatory sound* may keep the mating pair together for subsequent copulation.

6. *The recognition sound* has no proven function. Possibly it keeps groups or pairs of subsocial individuals together in burrows.

The first acoustical signal in the cricket system, produced perhaps 150 million years ago, must have been a soft sound that operated only between individuals in close proximity. Otherwise both the auditory organ and the signalling device would have had to appear suddenly, not only in complex form, but already tuned together—a possibility too remote to be worthy of serious consideration. The only soft, close-proximity signals among modern crickets are courtship sounds, and it is likely that this reproductive context was the one in which the first cricket chirp was produced. All the other signals are probably outgrowths of this fundamental situation.

The close functional relationships between courtship and calling suggest that the calling song is principally an absentee courtship signal; it arose as a result of an original courtship signal becoming more intense and of longer duration, finally being produced in the absence of the female and attracting her without the tactual and chemical signals usually present during courtship. Special aggressive signals probably arose as modifications of the calling song after it had already become a mediator of male-male interactions, as well as of male-female interactions. There seem to be two ways in which such duality in function could have developed. One is by having the same structural sound units affect two different kinds of individuals differently (here, the male and the female). The other is by developing two separate components in the signal, one an aggressive effect for other males, and the other a calling effect for females. The two different components can then be produced alternately during singing. Only a few crickets in Africa and Australia, among those studied, seem to have taken the second alternative, although many long-horned grasshoppers, katydids, and cicadas have done so.

Postcopulatory signals appear to have evolved from courtship singing in tree crickets. The female tree cricket stays on the male's back after the initial copulation and feeds on the secretions of a gland under the uplifted wings. Postcopulatory signals in the few field crickets that have them have evidently evolved from the calling sound. Postcopulatory sounds in these two cases are still similar to the courtship and calling sounds, respectively. Only a few crickets have postcopulatory sounds; it surprised us when we first saw a male cricket singing right after copulating, since the usual field cricket male cannot call again until he has developed another spermatophore, or sperm sac, and is ready to copulate again. That a single sound can function in these two different contexts, calling (pair forming) and postcopulatory, may result from the great difference between the two situations, which reduces the likelihood of confusion.

The so-called recognition signal of crickets is too poorly understood for much speculation about its origin, but where it occurs it seems structurally similar to courtship signals. This signal is made only by burrowing crickets that have developed elaborate parental behavior, and it is the only one supposedly produced by females as well as males. This suggests that its function may be interchangeable between the sexes.

Perhaps the most significant pressure for evolutionary change in cricket songs is the advantage of distinctiveness in the calling signals of species that are reproductively active in the same places at the same times. As many as thirty or forty cricket species may be calling together in a single habitat in North America, and if a female could not distinguish the sounds of her own males, her performance would be inefficient, to say the least. As we might expect, no two species that live together in this way have the same songs. Furthermore, some closely related species that do not live together, because they are either seasonally or geographically isolated, do have the same calling songs—in fact, almost exactly the same repertoires!

The consistent song differences between species that live together provide an extraordinarily powerful tool for the taxonomist. Using song as his single initial clue, he can obtain specimens of every species in an area within a short time. As a result, a great many puzzles in distribution patterns, morphological variation, and complexities in life histories are being solved, and some hope can be held out for accurate recognition of all cricket species in the near future, at least in the regions where field study and behavioral work can be carried out.

The most important species differences in cricket songs are in their rhythm patterns. In some cases, the stridulatory rhythms have become complex, involving pulse pairing, gradual increases and decreases in the rate of pulse delivery, and progressive, program-like changes requiring a minute or more for completion.

Teleogryllus commodus has one of the most complex repertoires known among crickets. Its calling song is a remarkable alternation of chirps and trills, pleasant to the human ear. Among the various species from Africa, South America, Hawaii, Jamaica, and other exotic places that are continually singing in my laboratory, I particularly enjoy listening to this Australian species and reflecting that this must have been one of the loudest and most persistent sounds in the environment of the Australian aborigines all during their evolution.

European man had the fairly simple chirps of the European field cricket, *Gryllus campestris,* and the house cricket, *Acheta domesticus;* the American Indians had a whole array of chirping and trilling species. It is not difficult at night, with the light off in my laboratory—where there are often forty or fifty species chirping and trilling together—for me to close my eyes and imagine myself alone in some primeval habitat thousands of years prior to the advent of civilization, surrounded, as men were then more than now, by the cheery bedlam of countless crickets chirping messages more than a hundred million years old.

11

Evolution of Nest Building

NICHOLAS E. COLLIAS

The lives of birds usually center about the nest during the most active part of their existence. The importance of the study of nests was recognized by ornithologists of a past generation in the creation of a special term, "caliology" (from Greek *kalia,* hut or nest, + ology), for this area of scientific investigation. As a rule, however, nest collections of most museums compare very poorly indeed with the collections of bird skins to be found in the same institutions.

With the growing emphasis on the study of the living bird, there is a renewed emphasis both on the study of nests and on the behavior patterns associated with their construction and use. There is also an increasing recognition that nests and nest sites often bring to a focus the principal habitat requirements of a species. The extreme variations in nest form, structure, and elaborateness possess a significance that, to be understood, often requires close study of the bird and its nest under natural conditions. Furthermore, like other characteristics of species, the type of nest built depends on an evolutionary history to which we may gain some clue by means of comparative study of related species. A nest may be defined as an external structure that contains eggs and young and that aids in their survival and growth.

Building a nest often requires a great deal of time and energy in many species of birds. It is common for birds to make a thousand trips or more to gather and transport all the necessary materials. Natural selection may be expected to favor any behavior patterns that economize on undue effort, provided that some crucial advantage of the species is not thereby sacrificed. For example, theft of nest materials from other members of the species is very common among birds, but it is obvious that undue use of this method by all the individuals of a species could have a serious effect on the reproductive rate at the population level. For example, years ago Alexander Skutch estimated that up to half the nesting failures he observed in a population of Rieffer's hummingbird *(Amazilia tzacatl)* in Central America resulted from stealing of nest materials from nests of other hummingbirds, resulting in the collapse of nests with eggs or nestlings. It is evident that the necessity to protect the nest itself has been an important force in the evolution of what is commonly called territorial behavior.

Competition between closely related species often results in the evolution of great differences in habitats and nest sites—differences that may give the birds their names. Compare, for instance, barn, cliff, cave, tree, and bank swallows, all well-known species in North America. In turn, differences in the nature of the substrate for the nest imposes special engineering requirements with regard to materials, form, structure, and placement of the nest for each species.

The primary and general functions of a bird's nest are to help insure warmth and safety for the developing eggs and young, but if the evolutionary forces involved in nest-building characteristics of any avian species are to be fully clarified, one must often be familiar with other aspects of its life history as well. Convergence in nest form and structure between unrelated species often furnishes significant clues to the nature of the ecological forces operating in the evolution of the characteristic nest.

My main emphasis is not to use nest characters to help develop particular phylogenies in groups of related species, but rather to attempt to show how one can analyze the ecological nature of the selection pressures that have led to the evolution of the main types of nests. Of course, the problems of warmth and safety for the young are generally most acute for small birds and their young, which explains why, as a rule, birds of small body size build nests that are more elaborate and better concealed than are those of larger birds.

When birds evolved from reptiles they probably developed the ability to maintain a high, constant body temperature, more or less coincidental with the ability to fly. It seems likely that during the transitional evolutionary period, when flying ability and temperature regulatory mechanisms were being perfected, many birds became torpid during very cool nights. A few birds known today, such as the poor-will *(Phalaenoptilus nuttallii)*, become torpid during cold weather. The probable imperfection of body temperature control in ancestral birds is an argument in favor of the theory that the first birds did not incubate their eggs by sitting on them, as do most modern birds, but probably adopted some other means. Perhaps they buried their eggs in the soil and relied on heat furnished either by decaying vegetation or the sun. Many reptiles do this, as do birds of the family Megapodiidae. Within the confines of one genus, *Megapodius,* the nest may vary from a simple, small pit dug in the sand, large enough for just one egg, to gigantic mounds of soil and decaying vegetation from 30 to 60 feet long and reaching 15 feet in height—the largest bird nests known. Some of the megapodes have developed an efficient control over the temperature in their nest mounds far beyond that seen in any reptiles. One megapode, the mallee fowl *(Leipoa ocellata)*, lives in arid regions of Australia where temperatures vary from below zero to above 38°C (100.4°F), and where even in midsummer the nighttime temperature may be 17° cooler than the day temperature. Yet, by varying the depth of the soil over the eggs and thus the degree of insulation from cold or of exposure to the sun, this bird maintains a relatively constant incubation temperature of between 32°C (89.6°F) and 35°C (95°F), as H. J. Frith discovered by making careful measurements throughout the breeding season.

At the close of the Mesozoic, the climate changed from humid tropical or subtropical to drier conditions, with greater extremes in temperature. This was probably met in early avian evolution by two different solutions to egg incubation. Some birds were or became mound builders, and evolved considerable efficiency in regulating the temperature in the mound about the eggs. Other birds had or developed the method of incubation by application of parental body heat to the egg.

Once birds had evolved the ability to maintain a high body temperature throughout the night, a strong selection pressure for direct parental incubation of the eggs would be established coincidentally. Eggs of reptiles often take months to hatch, whereas those of birds frequently hatch within a matter of weeks. Total predation on the eggs would diminish as the developmental period was shortened. The danger of predation from various nocturnal enemies, especially from the small contemporary mammals, would favor the habit of staying with the eggs and defending them, if necessary, during the night. F. H. Herrick has suggested that the origin of incubation by sitting on the eggs probably arose from the tendency of birds to conceal them as a protection from potential predators.

Use of natural or excavated cavities is common among birds and, in some instances, has evolved into quite elaborate excavations in the ground, in banks, or in trees. Almost half the orders of birds recognized by Ernst Mayr and Dean Amadon in their classification of the birds of the world contain some species that nest in cavities. Whole orders of cavity nesters are represented by the kiwis, parrots, trogons, coraciiform (kingfishers and their relatives), and piciform (woodpeckers and their relatives) birds. The habit of nesting in cavities furnishes considerable shelter and safety, particularly to small birds. Populations of house wrens *(Troglodytes aëdon)*, for instance, have often been increased by putting out a good supply of nest boxes. The shelter and safety of cavities has resulted in intense competition for these nesting sites. The house wren discourages competition by other birds such as the prothonotary warbler *(Protonotaria citrea)* by puncturing the warbler's eggs. Along with each size class of woodpecker goes a host of other species that compete with the woodpeckers for the corresponding size of nest cavity. The European starling is notorious in this regard. I have seen a starling in Ohio seize a yellow-shafted flicker by the tail and cast it out of the flicker's freshly dug tree hole in which a pair of starlings subsequently reared a brood. The German nature photographer Heinz Sielmann observed that when a European nuthatch *(Sitta europaea)* takes over a tree cavity it forestalls its chief rivals, the starlings, by collecting mud from nearby puddles and plastering it around the entrance to the tree hole until the entrance is so small and narrow that, while the nuthatch can slip through, the starling cannot.

Different stages in the evolution of nest sites in tree holes are represented by different species. Some use natural cavities, others modify these cavities or excavate cavities in soft or decaying wood or even in hard, living trees, as do some of the larger woodpeckers. Similarly, in the case of birds that nest in holes in the ground, various degrees of specialization are illustrated by different species. These range from a shallow scrape in many ground nesters, to a relatively short burrow like that of the rough-winged swallow *(Stelgidopteryx ruficollis)*, to larger burrows such as those made by the related bank swallow *(Riparia riparia)*, which may be six or more feet long, tunneled in something of an upward course and providing protection against driving rain.

The evolutionary climax of excavated nests is the construction of nesting cavities by certain birds inside the nests of social insects. V. A. Hindwood has listed forty-nine species of birds, including kingfishers, parrots, trogons, puffbirds, jacamars, and a cotinga, that are known to breed in termite nests. In fact, some 25 percent of the species of kingfishers of the world nest in termitaria. As the excavation by the bird progresses, the termites seal the exposed portions of their nest so there is no actual contact between birds and insects. Many birds that breed in termite nests do not normally eat the insects of the colony in which they are nesting. Birds that breed in nests of social insects are all from taxonomic groups characterized by nesting in cavities, and cavities in old and deserted termite nests are at times used by birds that normally breed in earth banks or tree holes, suggesting how the habit might have evolved.

Nesting in a hole goes a long way toward meeting the essential nest functions of warmth and safety, and thereby actually tends to block further evolution of truly elaborate increment nests built up from specific materials. In fact, nests built within cavities may undergo a regressive evolution. One can see all degrees of increasing simplification and reduction from an elaborate roofed nest to a mere pad, as in the case of the Old World sparrows (Passerinae) that nest in tree holes.

In contrast to cavity nesters, birds that build open nests on the ground are subject to a greater chance of nest failure. Consequently, there is a strong selection pressure to build an adequate nest, to develop other special means of parental care, or to evolve markedly efficient concealing coloration. Although with the origin of direct parental incubation it was no longer necessary to dig a pit for the eggs, most birds that nest on the surface of the ground today still begin their nest by making a circular scrape with the feet while crouching low and turning in different directions. This hollow may then be lined with various materials to protect the eggs from the cold, damp ground, while a rim of materials around the body of the sitting parent provides added insulation for the eggs. The materials are pushed to the periphery of this rim and built up into a circular form by much the same sort of movements of the feet and body as are involved in making the initial scrape in the ground. Many ground-nesting birds prevent the flattening down of the peripheral raised rim of the nest by repeatedly reaching out with the bill and drawing materials in to the breast or passing them back along one side of the body before dropping them. These patterns of making a nest can be seen, for example, in the Canada goose.

Nests built on the surface of the ground are especially liable to be flooded. These nests are often built on slight elevations and, as in the case of the Canada goose, may be built up higher during a flood. The painted snipe *(Rostratula benghalensis)* in Australia may lay its eggs on the bare ground when the earth is dry, but if it is covered with water, a solid nest of rushes and herbage is made. Similarly, the Adelie penguin *(Pygoscelis adeliae)* builds up its nest of small stones if thaws cause flooding. In the Antarctic, where this penguin nests, William Sladen noticed one nest that had a stream of ice-cold water running through it. The male on the nest kept reaching forward, collecting and arranging stones about himself and his half-submerged eggs. By the next day the nest and eggs were above water, and eventually the eggs hatched.

Parental behavior may supplement or even substitute entirely for a nest under severe environmental conditions. In the Arctic, persistent close incubation by the parent bird is necessary and characteristic, regardless of whether or not the nest is well insulated. It has been found that the semipalmated sandpiper *(Ereunetes pusillus)*, which builds no nest, keeps its eggs as warm as do other Arctic species that have substantial nests. The emperor penguin *(Aptenodytes fosteri)*, which breeds in the Antarctic winter, has no nest, but rests its single egg on the feet, covers it with a fold of skin from the abdomen, and incubates it against the body. Probably no other animal breeds under such trying conditions. At an opposite extreme, eggs or nestlings exposed to strong tropical or subtropical sun in open situations are customarily shaded by the body and wings of the parent, as in the case of the sooty tern *(Sterna fuscata)* of Midway Island, whose nest is a mere scrape in the coral sand.

Birds nesting on the ground are subject to high predation, and it is no accident that the classical cases of concealing coloration are found in birds, like the ptarmigan, which lay their eggs in an open nest on the ground. In certain cases, the color pattern of the eggs and the young, as in the European stone curlew *(Burhinus oedicnemus)*, or of the young and the parent, as in the whippoorwill *(Antrostomus vociferus)*, matches the surroundings so closely that the nest has disappeared in evolution, presumably because a nest itself would attract attention and be too conspicuous.

The dangers of ground nesting and the intense competition for tree holes have apparently provided a strong selection pressure for the evolution of increment nests placed on the branches of trees or, in some cases, against the faces of cliffs. Species of birds with precocial young (covered with down and able to move about) are generally ground nesters, whereas species with altricial (naked and helpless) young frequently nest on trees or bushes. In the prairie country of northwestern Oklahoma there are few trees, and R. L. Downing found that those mourning doves *(Zenaidura macroura)* that nested in trees were almost twice as successful in fledging young as were those that nested on the ground. There was a definite preference among the doves here for nesting in trees. In fact, Mrs. Margaret Nice, also in Oklahoma, found long ago that pairs of mourning doves nesting within forks of trees had a greater success than did those pairs nesting farther out on branches. Conversely, under safe nesting conditions, certain species of birds (such as the osprey and robin on Gardiner's Island, New York) that normally nest in trees may nest on the ground, thereby conserving the energy required to fly up into a tree with nest materials or with food for nestlings.

Tree nesting requires the solution of new types of engineering problems. The nature of the materials used varies with the body size of the bird and its lifting power. Large birds use large twigs and even branches, which will not readily be blown out of trees by the wind. Medium-sized birds use small twigs or grasses or both, sometimes adding mud to help attach and bind the materials. Many small birds use spider or insect silk as a binding material for the attachment of the nest to the substrate and to bind various other materials of the nest together.

The platform nests of large birds, such as the American bald eagle *(Haliaeetus leucocephalus)* and the European white stork *(Ciconia ciconia)*, may have twigs and branches added year after year and may become very large and very old. F. H. Herrick describes an eagle nest 12 feet tall and 81/2 feet across, and estimates that it weighed over two tons when, in its thirty-sixth year, it fell during a storm, together with the tree. F. Haverschmidt has managed to date back to 1549 one white stork nest that was still in use in 1930.

It seems probable that every type of material characteristic of the nest of a given species has a definite function, and that the proportions of different types of materials used vary not only with availability but also with the requirements of particular substrate and habitat conditions. Otto Horváth observed that robins' nests in British Columbia contained more mud when the birds had to use short building materials, more tough and flexible rootlets when the nest was in an especially windy spot, and more moss when it was in a relatively cold microclimate.

Cup nests of very small birds are likely to be heavily insulated, as is true of the nests of most species of hummingbirds. There is some evidence that, compared with lowland species, hummingbirds that nest in high mountains build nests with relatively thick walls or seek the protection of caves.

Nests attached to the vertical faces of cliffs, caves, or buildings furnish protection against nonavian predators, but pose special problems for attachment of the nest. The swifts have generally used adhesive saliva, while the swallows have evolved toward more frequent use of mud, probably with some admixture of saliva. Different species of swiftlets *(Collocalia)* can be arranged in a graded series from such species as *C. francica,* which make nests using pure saliva (the source of the ideal bird's-nest-soup of the Chinese), through other species that use various admixtures of plant materials, to those that build nests of more conventional types. The nest cement of *C. fuciphaga* is sparse and soft, and the nest, which is composed largely of moss and other plant materials, can only be placed on an irregularity in the cave that will take all or a good part of the weight of the nest. In contrast, nests of other cave swiftlets can be glued to vertical walls in the cave.

Building of a roofed increment nest is very rare among nonpasserine birds, whereas almost half of eighty-two families and distinctive subfamilies of passerine birds recognized by Mayr and Amadon construct roofed nests or contain species that do so. Although roofed nests, aside from use of natural cavities, are unusual among passerine birds of the North Temperate Zone, they are typical of many tropical genera and families of passerines. In most instances, birds having nonpensile roofed nests first make a basal platform and then build up the sides and roof, and this sequence of building suggests that roofed nests probably evolved from nests that were at one time open above.

Roofed nests may be made of very different materials by different birds—

woven or thatched of grasses in many weaverbirds, of plant fibers in certain icterids, of short, heterogenous plant materials bound by spider silk in sunbirds and some titmice, and of mud in cliff swallows. The convergent evolution in such diverse instances is evidence of the great importance of a roof in the life of small nesting birds.

The roof of domed nests is important in shading young birds from the sun. Solar radiation is most intense in the tropics and would quickly kill small, naked altricial nestlings exposed to direct rays. The Galapagos finches (Geospizinae), for instance, have an equatorial habitat and, unlike most Fringillidae, build roofed nests.

One function of the roof of domed nests must be to shed rain. Most small birds in the tropics nest during the rainy season when insect food is abundant. Skitch observed that the nests of the yellow-rumped cacique *(Cacicus cela)* in Central America are all open at the top during the early part of the breeding season before the rains. But as the rains begin, after the eggs have been laid, the top of the entrance is gradually roofed over and the nest entrance becomes a bent tube opening downward.

Protection from bird and mammal predators is aided by placement of nests in dense cover, especially in thorn trees. The buffalo weavers of Africa build a thorny cover or shell to the nest. Furthermore, the whiteheaded buffalo weaver *(Dinemellia)* is famous for placing thorny twigs along the boughs leading to its nest.

Among the predators of nestlings, snakes are more numerous and varied in the tropics than in colder localities, and perhaps roofed nests help deter snakes as well as other enemies. The weaverbirds all build domed nests, and, in addition, the nests placed in trees tend to evolve a firm pensile attachment and a bottom entrance with a long entrance tube, enhancing protection from snakes. In East Africa, Van Someren once watched a green tree-snake trying to get at the young in a spectacled weaver's nest. The snake negotiated the slender, pendent branch and reached the nest, but could not manage the twelve-inch tubular entrance and fell into the pond below.

A woven construction facilitates the evolution of roofed and pendulous nests and enhances the nest's coherence. A whole series of representative stages can be traced in the weaverbirds from loose, crude, irregular weaving to the close, neat, and regular pattern to be found especially in those species that build pendulous nests with long entrance tubes. The nest of Cassin's malimbe *(Malimbus cassini)*, a black and red forest weaver of Central Africa, is perhaps the most skillfully constructed nest made by any bird.

In contrast to its importance among the social insects, a compound nest has been evolved by only a few species of birds. The compound nest consists of a common nest mass in which more than one pair of birds or more than one female of the same species occupy separate compartments. Such birds include the palm chat *(Dulus dominicus)* of Haiti, the monk parakeet *(Myiopsitta monacha)* of Argentina, the black buffalo weaver *(Bubalornis)* of Africa, and the sociable weaver *(Philetairus socius)* of the Kalahari Desert in South Africa.

The nest masses of the sociable weaver have often been compared to haystacks in a tree. These nests are not woven, but are thatched with dry grass stems. Each nest mass is often several feet thick, of irregular extent, and may be over fifteen feet long in the longest dimension. The top of each nest mass is dome-shaped, the underside relatively flat and riddled with scores of separate nest chambers. Many different individuals may work together on the common roof, which may be one key to the evolution of this remarkable structure. The roof enhances protection from predation for all, as does the outer thorny shell in nests of the black buffalo weaver or the projecting eaves in those of the monk parakeet.

Special security from predation seems to be an important factor in making possible gregarious breeding, a phenomenon that is rather rare among small land birds, although common in sea birds on remote islets or inaccessible cliffs. One well-known example of gregarious nesting among passerine birds in an obviously safe nest site is that of the cliff swallows, whose mud nests are placed on the vertical face of a cliff or building. Colonies of the sociable weaver are frequently found, in camel's-thorn acacia trees, in which each of the many thorns may contain a colony of ants *(Natural History,* January 1965). The habit of nesting in association with colonies of noxious insects is found in many African birds, and may have been a predisposing force in the evolution of the sociable weaver's compound nest. Possible intermediate stages are represented by the nests of a related species in East Africa, the grey-headed sociable weaver *(Pseudonigrita arnaudi).* This gregarious species breeds in small colonies in which the different nests may be well separated, but when placed in ant-gall acacias, whose thorns give added security, many of the nests are grouped into common masses. We have counted up to nine nests in one mass.

When one considers the extent of variation in birds' nests there seems a vast difference between, say, the shallow scrape of a sooty tern in the coral sand of a tropical island and the immense communal dwellings of the little sociable weaver or the exquisitely woven cradle of Cassin's malimbe. These variations are an exciting challenge to the ornithologist who would attempt to explain their origin in terms of the complex ecological and behavioral forces that have shaped the nests of birds in the course of evolution.

PART 4
BIOGEOGRAPHY AND EVOLUTION

The geographic distribution of animals and plants has an old, and straightforward, connection with the very idea of evolution. If all organisms had been created in the Garden of Eden, later to be collected by Noah and subsequently released from Mt. Ararat, we might expect the greatest concentration of life to center around the Middle East, with life fanning out from there, becoming less dense and diverse the further we sample from this center of origin and dispersal. In any case, we would not expect to travel the world over and find highly localized flora and fauna that bear no particularly close resemblance to the denizens of the Ark.

Darwin realized that the highly diverse nature of life spread out over the globe, a geographically based diversity that by his day was becoming increasingly appreciated, afforded a powerful boost to the argument that life has had a long and complex history. That similar species seemed to replace one another from region to region led to the notion of "ecological vicars," an idea closely related to interspecific competition and mutual exclusion of closely related species. And the concept of allopatric speciation, long the dominant model of the speciation process, casts geography in the leading role in the differentiation and origin of new species.

Geographic isolation allows elements of the local biota to go their own separate evolutionary ways. The pattern is strongly hierarchical: each major biogeographic region has many faunal and floral entities unique (endemic) to it. As we shall see in some of the following essays, entire families, orders and, in some cases, taxa of even higher rank may be restricted to particular regions. But within any biogeographic area, there are inevitably subregions of endemism. It is the task of all biogeographers to work out the distributions of taxa of an area, and to come to grips with the history of those distributions.

The essays in part 4 cover several approaches to historical biogeography. Analysts of living organisms frequently take a functionalist approach, seeing distributions governed by an interplay of the biological requirements of the organisms and the physical circumstances in which they find themselves. Often their explanations assume a center of origin for a group, with subsequent dispersal leading to patterns of distribution in the present day. This is the central thesis of the first essay, by James Hanken, James F. Lynch, and David B. Wake. They depict the diversification of lungless salamanders in the New World tropics as an adaptive radiation that followed successful dispersal from the ancient Appalachian highland of eastern North America. Thus successful invasion of fresh territories may lead to fresh evolutionary opportunities. Note, too, how Hanken et al. rely on the notion of interspecific competition as an organizing force in lungless salamander communities—harking back to Wiens' essay, and taking a rather different view of the matter.

Francois Vuilleumier, an ornithologist on the staff of the American Museum of Natural History, also takes the distribution of living organisms as his point of departure. This time it is bird faunas of the Andes. In the north, the paramos form a series of islands separated by unsuitable lowland habitat. In keeping with the hierarchic structure typical of biogeographic regions, Vuilleumier finds a second area to the south *(puna)* with an avian fauna similar to that of the paramos. Vuilleumier takes a two-pronged approach to his study. Using a statistical analysis of faunal resemblance, he invokes both local geologic and climatologic history with dispersal as a means of explaining the patterns of similarity he sees within and between the two major regions. In particular, oscillations of both glaciation and mountain building have changed the geography radically, and set the stage for speciation to occur (returning to the theme of part 2). Also, with Wiens, Vuilleumier has his doubts on the importance of interspecific competition as an important determinant of local bird distributions. His essay, with its cautious conclusions, is a good example of how the data of a variety of scientific disciplines may all be brought to bear on biogeographic problems.

With Edwin H. Colbert, Curator Emeritus at the American Museum, we shift focus from the Recent to the fossilized, distant past. In so doing, we enter a different conceptual world, if for no other reason than that paleontologists have a great deal of difficulty formulating rigorous hypotheses in functional terms. We can only speculate in highly generalized terms about past physiologies and adaptations; and our ability to reconstruct past environments is likewise severely limited. Thus we paleontologists tend to prefer more purely historical explanations—which are really functional explanations at the coarse scale appropriate to our data. Thus our obsession with continental drift, or "plate tectonics," as the modern notion is known.

Colbert gives us a fascinating glimpse of dispersalist explanations in historical biogeography. When geologists firmly proclaimed the stability of continents, indeed there was no choice other than the mental construction of land bridges to explain the distributions of land plants and animals between continents. But there is a simpler explanation for much of the curious biotic resemblance among portions of the continents of Antarctica, Australia, Africa, South America and India: all were parts of the ancient and now fragmented supercontinent of Gondwana. Colbert wrote this article at a time, as he says, when "drift" was just in the throes of becoming an orthodox theory of the history of the earth—hence his desire to include paleo-biogeographic data as part of the evidence favoring the idea of the past motion of continents. But the main effect of plate tectonics on historical, evolutionary biogeography has been to give us insight into a powerful physical force that shapes the distributions of animals and plants. Vicariance biogeography, which emphasizes the fragmentation of biotas rather than random dispersal of organisms as the main source of biogeographic patterns, dovetails nicely with this notion: relative plate motions modify distributions and yield more regular patterns of endemism and similarity between regions than haphazard dispersal is likely to do. Colbert shows us how well the idea works to explain the similar ancient faunas found on continents now far removed from one another.

Sticking with Gondwana, but shifting back still farther in time, I wrote an analysis of the evolutionary history of the Calmoniidae, a family of trilobites restricted to that region in Upper Silurian and Lower Devonian times. Several evolutionary themes are featured here, including adaptive radiations, convergent evolution, and speciation. I reiterate what has become one of the major themes in evolutionary paleontology in recent years, one documented in greater detail in the essay on punctuated equilibria reprinted in part 2: stasis, or evolutionary stability within species. For all of these patterns, geography seems to be a key underlying factor. And, once again, the biogeographic patterns within Devonian Gondwana are clearly hierarchical. There is, first, the entire Malvino-Kaffric faunal province, distinctly set apart from the rest of world. Secondly, the province is fragmented into three major regions; and, thirdly, patterns of geographic variation and speciation more locally break the region down into still smaller bits and pieces. The interface between geography and evolution is indeed multifarious.

The final essay is really a codicil to my preceding article. Offering a glimpse into the workaday scientific world, it addresses the first level of Gondwanan biogeography—the outer limits of that ancient Devonian faunal province—and also shows how fossils, at least *potentially,* can be used to identify wandering pieces of ancient continents, picking up a bit on Colbert's theme: fossils can be informative to geology just as geologic history can help us understand the evolutionary and biogeographic histories of organisms.

12

Salamander Invasion of the Tropics

JAMES HANKEN, JAMES F. LYNCH,
and
DAVID B. WAKE

What is evolutionary success? Perhaps the most obvious example is an adaptive radiation: the divergence of members of a single lineage into a number of different ecological niches or adaptive roles. This divergence can involve changes of morphology, physiology, ecology, and behavior. By investigating adaptive radiations, we can learn the causes and limitations of evolutionary change.

In many ways, salamanders might seem unlikely subjects for the study of adaptive radiation. As is typical of most amphibians, salamanders must maintain a moist skin and are frequently restricted to aquatic sites in or near ponds, streams, and seepages or under moist logs or leaf litter in humid forests. Because salamanders are ectothermic (cold-blooded), their energetic needs are slight—usually only a fraction of those of birds and mammals. They can feed infrequently and remain inactive for long periods; this, plus their usually small size, makes them inconspicuous. For these reasons, some biologists have dismissed them as rare and of little ecological significance.

Despite their apparent low profile, in many habitats salamanders are the most abundant vertebrates. For example, salamanders in a New Hampshire forest were found to exceed birds and mammals in both numbers and biomass. Further, salamanders display a variety of body shapes and sizes, ranging from inch-long species to an Asian genus whose members attain lengths of five feet. Some salamanders are fully aquatic, others are partly so, and many are fully terrestrial. Most species have well-developed legs and feet, but some groups have tiny limbs, and others lack hind limbs entirely. Some salamanders are subterranean, while others spend their entire lives far above the ground in the canopy of tropical forests.

The approximately 325 species of living salamanders are divided among nine families. However, more than 200 species belong to one family, the Plethodontidae, or lungless salamanders, and about 150 of these, or nearly half of all salamanders, are members of one subgroup, the tribe Bolitoglossini, or, literally translated from Greek, the "mushroom-tongued" salamanders. Salamanders as a group arose in the Northern Hemisphere, and virtually all living nonbolitoglossine salamanders are confined to the north temperate portions of Asia, Europe, and North America. In contrast, all but about a dozen bolitoglossines live at tropical latitudes in Middle and South America. How can we account for this successful invasion of the tropics?

The earliest lungless salamanders were descendants of stream-dwelling salamanders of the ancient Appalachian highlands of eastern North America. The most primitive living lungless salamanders are still found in stream habitats in the Appalachians where they maintain a 100-million-year-old life history pattern. Semiaquatic adults mate and deposit eggs at streamside. Carnivorous aquatic larvae that emerge from these eggs later metamorphose into terrestrially adapted adults that leave the water. This reproductive pattern, although typical of many other amphibians, constrains ecological and evolutionary diversification. Most important is the restriction of such amphibians to areas where aquatic breeding habitat is at least periodically available. Dependence on aquatic breeding sites also limits the ability of amphibians to disperse through areas without free surface water.

Exceptions to this life history pattern are rare in most salamander families. One alternative is demonstrated by a few species of the European genus *Salamandra*, in which females do not deposit eggs, but instead retain them in the oviduct where development proceeds. Young emerge either as well-developed aquatic larvae or fully metamorphosed terrestrial salamanders.

Many lungless salamanders, including all bolitoglossines, have evolved a different means of eliminating the restrictive aquatic larval stage. These salamanders retain the primitive pattern of laying eggs, but the eggs are deposited in moist, protected terrestrial sites where the female broods them for a long period (up to eight months in some species). At hatching, a fully functional salamander emerges, having achieved complete independence from aquatic habitats.

With this fully terrestrial reproductive mode, ancestral lungless salamanders dispersed across the humid temperate forests that covered most of North America during late Mesozoic and early Cenozoic times, ultimately extending west to the Pacific coast and south into the New World tropics. Subsequent episodes of mountain building and climatic change caused much of the temperate forest to disappear from the western half of North America in favor of more arid habitats, eliminating lungless salamanders from virtually all of the midcontinental region and northern Mexico.

Today, isolated survivors of these events give us some idea of the enormous and nearly continuous distribution that lungless salamanders must have enjoyed in the past. Relict species still may be found in pockets of favorable humid habitat in New Mexico, Oklahoma, Texas, and Arkansas, while in California's Mojave Desert, several distinctive species have recently been discovered leading precarious lives around tiny springs and seepages in otherwise uninhabitable mountains. Most startling is the present distribution of the genus *Hydromantes*, one of the two genera of nontropical bolitoglossines. The western North American ancestors of *Hydromantes* dispersed into Asia via the Bering land bridge across the northern Pacific Ocean and extended their range to the west and south until they reached the Mediterranean region. Today, we find two of the five species of *Hydromantes* inhabiting caves and crevices in limestone areas of northern Italy, southern France, and the island of Sardinia, while the remaining three species are restricted to the mountains of central and northern California.

A great expanse of inhospitable arid country at present separates the lungless salamanders of temperate North America from those of the tropics. The north-ernmost members of the tropical assemblage include several generalized species of the genera *Pseudoeurycea* and *Chiropterotriton* that live at high elevations in Mexico's northern mountains. In morphology and ecology, these animals resemble the presumed ancestors of all tropical salamanders. These ancestral forms contin-ued their southward invasion via mountainous dispersal routes, leaving descendants throughout all of Middle America and much of South America as far as Bolivia.

The adaptive radiation of salamanders in the tropics has produced more than 150 known species assigned to nine distinct genera. As many as sixteen species may inhabit a single tropical mountain. Salamanders in the tropics are perhaps most abundant in mountainous areas, but they also are well represented at inter-mediate and low elevations, with the proportion of lowland species increasing southward. In southern Mexico and Guatemala more than 40 percent of the species are found, at least in part, below 5000 feet and 14 percent occur below 1500 feet. Farther south in Costa Rica, Panama, and northwestern South America, more than 85 percent of the species occur below 5000 feet, and fully a third of the species live below 1500 feet.

This information will help dispel the belief that tropical salamanders are restricted to high, mountainous areas with climates and habitats similar to cool, temperate forests. Salamanders are both diverse and locally common in the lowland forest where they thrive in conditions that are, by anyone's standards, truly tropical. Temperatures are characteristically high and rainfall may be highly seasonal, limiting activity by salamanders to only a portion of the year.

Although ecologists argue over the causes and effects of tropical diversity, a given tropical habitat will generally have more predators, parasites, and competitors than a corresponding temperate habitat. By conducting field studies we have attempted to understand how tropical salamanders have not only managed to invade tropical communities but to flourish in them. One important aspect concerns community structure. In North America, coexistence of similar species is achieved by a number of mechanisms. For example, in salamander-rich areas of the Appalachians, the species in a local community are about equally divided between aquatic and terrestrial forms. In species with aquatic reproduction, larval development may occur in springs, streams, or ponds, and breeding seasons may differ from species to species. Thus, both spatial and temporal overlap of larvae and, presumably, larval competition are reduced. Species that occur in the same habitat at the same time frequently differ in body size, and this is correlated with differences in the kinds and size of animal prey taken. All of these mechanisms may reduce competition between coexisting species.

Tropical salamander communities are different. As we have noted, all tropical salamanders are fully terrestrial, so separation along an aquatic–terrestrial gradient is not possible. Instead, two types of vertical segregation are especially important.

First, each tropical species has a narrow, precisely defined altitudinal distribution that frequently matches the complex, yet regular, elevational zonation of tropical plant communities. In this way, species present on the same mountain may actually never occur together because of non-overlapping elevational distributions. Although altitudinal layering also occurs in temperate regions, it is never as pronounced as in tropical habitats.

At any given elevation and habitat, further sorting is achieved by different species living at different heights above or below the forest floor. Some burrow, others confine their activity to the ground surface, while many literally take to the trees where epiphytic plants, especially bromeliads, are favored retreats. Lacking a functional root system, bromeliads attach themselves to the trunks and limbs of trees and obtain necessary moisture and nutrients from the atmosphere. Like salamanders, bromeliads thrive in areas with excessive humidity, particularly the cool, fogbound cloud forest that cloaks the slopes of many tropical mountains. There, a diverse community of salamanders and other small animals has evolved in the aerial swamp created by the upward-directed, tightly overlapping leaves of bromeliads. The association between salamanders and bromeliads is hardly casual.

Some salamander species occur nowhere else, and local densities can be great—as many as thirty-four salamanders have been found inside a single bromeliad in a Mexican cloud forest. We have found salamanders in bromeliads as high in trees as we have been able to search, and examination of recently felled trees (an all-too-common sight in the tropics) has revealed salamanders at least 100 feet above the ground.

Distinctive habits and habitat preferences of salamanders are often associated with morphological specialization, and several species, or even whole genera, are highly modified to survive in tropical environments. Arboreal montane species usually have small, flattened bodies that allow them to squeeze between tightly overlapping bromeliad leaves. Such species also possess a moderately prehensile tail and partially webbed hands and feet that adhere readily to wet leaf surfaces, providing great climbing ability. Burrowing forms have tiny limbs and extremely elongate bodies and tails that almost make them vertebrate worms. Members of the genera *Thorius* and *Parvimolge* are true miniatures. Living among leaf litter and bark on the forest floor, some adults may be only one inch in total length. This brings them very near the minimum size limit for vertebrates, and they possess various skeletal and sensory modifications to accommodate their reduced size.

The most specialized of all tropical salamanders live in the wet lowland forest. During the day most species conceal themselves in bark crevices, leaf axils, or other relatively inaccessible sites. Emerging only at night, they cling to a tree trunk, plant stem, or leaf surface in wait of prey. (This behavior explains why even experienced field biologists have tended to overlook lowland tropical salamanders.) Ground-dwelling salamanders are scarce in the lowland tropics; almost all lowland species are either arboreal or subterranean. Arboreal lowland salamanders are even more highly specialized than their relatives in the montane cloud forest. A fully prehensile tail and extensive webbing, which in effect turns the hands and feet into suction cups, allow a salamander to suspend itself, even upside down, from stems and leaves.

During the evolution of tropical salamanders, a premium has been placed on adaptations that reduce activity and make the most of available energy. A prime example involves prey capture. Most nontropical lungless salamanders ingest insects, millipedes, and snails by partially flipping out a sticky tongue, then retrieving it with the prey item attached. Simultaneously, the salamander may lunge forward and seize its victim in its jaws, especially when the prey is large. Because the front of the tongue is anchored to the lower jaw, the back of the tongue is flipped out at the prey. This resembles the movement of a clenched fist resting palm up on a table top as the fingers are first quickly outstretched and then reclenched. The anterior attachment of the tongue means that prey can be attacked only at close range.

In contrast to this primitive pattern, the tropical salamanders have evolved a highly protrusible tongue that can be fired accurately for long distances at astonishingly high speeds. Careful laboratory measurements have shown that in the genus *Bolitoglossa* the tongue can be extended to about a third of a salamander's body length and returned to the mouth with a captured prey in as few as ten milliseconds. The evolution of this efficient prey-capturing system has required a complete reorganization of the musculature and skeleton of the tongue to allow the mushroom-shaped tongue, now no longer attached in front, to be catapulted from the mouth. As an additional refinement, bolitoglossines have developed the ability to control the direction of tongue projection. Thus, the salamander need not directly face or actively pursue its prey. A highly modified tongue structure means that a salamander can instead wait for prey to approach within its increased firing range.

Regional species diversity of tropical salamanders is greatest in areas that are geologically most active. To fully understand the regional patterns of adaptation and diversity in salamanders, we must consider geologic factors as well as behavior and morphology. Throughout much of Middle and South America large–scale tectonic processes, ranging from massive volcanic activity to lateral slippage of continental plates, have created new habitat, destroyed previously habitable areas, erected barriers to dispersal, and fragmented once continuous salamander distributions.

A particularly good example of the connection between geologic events and biotic diversity and distribution is seen in the genus *Chiropterotriton*. Eight species of this group are found in west montane forests of Guatemala and adjacent Mexico and Honduras. Morphological, genetic, and ecological evidence agree that these species are more closely related to one another than to any other salamanders, insofar as they share a common ancestor. Each species is confined to its own mountain range or cluster of ranges, and no two species occur together. For tens of millions of years, this part of northern Central America has been the scene of violent geologic activity. The region straddles the juncture of three major crustal plates, and massive fault scarps crisscross a landscape dotted with some of the most spectacular volcanoes on earth. Originally continuous upland areas, some of which date from the early Cenozoic period, have been repeatedly fragmented by lateral faults, and large upland sections have been moved hundreds of miles from their original locations. Any ancestral *Chiropterotriton* species that inhabited the continuous upland habitat would have been fragmented into numerous subpopulations, each separated by impassible lowland barriers. Some of these isolated populations doubtless became extinct, but others continued to evolve independently, so that now each is sufficiently differentiated to qualify as a distinct species.

Why have salamanders experienced such a major adaptive radiation in the tropics? We can suggest the following explanations: (1) the prior evolution of a fully terrestrial life history preadapted bolitoglossine salamanders for life in the tropics, particularly in rugged, mountainous areas where free surface water is scarce; (2) limited activity, slow metabolism, and a simple, yet efficient, feeding mechanism allow salamanders to make slight energetic demands on an environment and invade and occupy otherwise closed tropical communities; (3) morphological specializations, including a diverity of locomotor patterns; (4) the availability of distinctly tropical microhabitats, especially arboreal bromeliads, enables more species to be packed into a given habitat; (5) the restriction of species distributions to the narrow elevational zones of climate and vegetation characteristic of tropical environments increases local diversity; and (6) the extraordinarily high intensity of geologic activity in Middle America and northern South America has promoted the formation of new species, thereby increasing regional diversity.

Tropical salamanders provide an unusually favorable opportunity to study patterns of evolution. Our effort has spanned more than a decade, and has at various times included the collaboration of anatomists, physiologists, taxonomists, ecologists, biochemists, and geneticists. It is sad to note that thoughtless devastation of fragile tropical habitats by humans threatens to obliterate many of the most interesting pages of this story before they can be read.

13

The Origin of High Andean Birds

FRANÇOIS VUILLEUMIER

Breathtakingly beautiful at times, with a palette of bright colors under the hot, midday sun, but forbidding at others, with strong winds, hailstorms, persistent fogs, and the numbing cold of early morning, the high-altitude realms of the Andes continue to draw me back. Since 1964 I have made six expeditions, totaling about twenty months, that have taken me all along the Andean backbone, from Venezuela to northern Patagonia. Why? Because the Andes are, as ornithologist Frank M. Chapman put it half a century ago, "a new world."

Geologically recent, this narrow, 4000-mile-long mountain chain provides the biogeographer with a wonderful setup for inquiries about fundamental evolutionary problems. The paramos of the Venezuelan and Colombian Andes—wet, grassy alpinelike areas extending from 10,500 feet up to the snowline—are islands of open vegetation separated from each other by lower-altitude cloud forests. Farther south, in Peru and Bolivia, the high-altitude zone, known as the *puna*, broadens into an arid and sparsely vegetated plateau stretching to the horizon. Not surprisingly, paramo and *puna* habitats have little in common with the steaming lowland or the cool montane forests found in other parts of South America. But they also share little with the hot, grassy savannas of Venezuela or the open shrublands of Brazil. The high Andes are more like the Patagonian steppes at the far south of the continent.

The resemblance between habitats of the high Andes and Patagonia applies to the bird fauna as well. An ornithologist familiar with Patagonian birds who was transported 1500 miles north to the Bolivian high plateau would feel at home. The two areas share a number of genera and several species; even the missing families are the same. How, then, did the high Andean birds evolve? Do they derive from Patagonia or do Patagonian birds derive from the high Andes? How much of the fauna is the result of speciation in the high Andes themselves and how much the result of immigration from lowland areas?

These questions boil down to a classic problem in biogeography; in this case, how to explain the history of about 180 species of birds. The procedure is a bit like detective work: look at the evidence, omit no detail, and reconstruct the circumstances of the crime. The crime was committed in only one way, but the circumstances can almost always be variously reconstructed, depending on the interpretation of certain crucial details. Similarly, although the fauna of a region evolved in only one way, different scientists working with the same biogeographical evidence may well suggest dissimilar reconstructions of that evolution.

We need two kinds of clues: historical ones—ideally, abundant fossil remains—and modern ones, revealed in the distribution patterns of living species. Fossils of high Andean birds are so scarce that they are virtually useless. Biogeographers are, however, beginning to learn about the history of some Andean environments through the work of palynologists, who study pollen, and about the history of the mountains themselves through the efforts of geologists.

In the Andes today, permanent glaciers are found only at the highest altitudes, but in the past they extended much farther down the mountainsides. In several areas of Venezuela, Colombia, and Peru, moraines and fluvioglacial accumulations of rocks and other materials attest to between one and three major glacial advances. But if the Andes, like parts of Europe and North America, witnessed four major advances, why are they not all recorded? One view is that at the time of the earliest glacial advances in the Northern Hemisphere, the Andes were not sufficiently high to have had glaciers. The high, alpinelike environments of the Andes (the wet paramos of Venezuela, Colombia, and Ecuador and the drier *puna* of Peru, Bolivia, northern Argentina, and northern Chile) probably only date from the mid- to late-Pleistocene era, one to two million years ago.

While not definitive, pollen data also suggest that the Andes have been high for a geologically short time. Palynologists have carried out several detailed studies of cores taken from ancient bogs in the Andes of Venezuela, Colombia, and Bolivia. They have dated the peat from different core depths and counted and identified by species the pollen grains found in the peat. The resultant "pollen profiles" indicate how much pollen was contributed by different species of trees or grasses at different times and different places. For example, at one locality in the Colombian Andes, currently at an altitude of 13,500 feet and situated fully within the paramo zone, the vegetation apparently changed repeatedly during the last 15,000 to 20,000 years. At times, woodlands occurred at the site, while in other periods the vegetation resembled the wet grassland that now dominates the area. The overall picture of the Andes is one of vegetation zones that have been alternately lower and higher over about the past 100,000 years. During cold, glacial episodes, alpinelike vegetation reached much lower altitudes than it does now. By contrast, at the peak of warmer, interglacial episodes, the zones moved up, so that the paramo and *puna* region was even more restricted to mountaintops than today.

If a given bird species now found exclusively in alpinelike grassland had the same habitat requirements in the past, its distribution would have changed over time according to the periodic contractions and expansions of its environment. Theoretically, this "cycle" of habitat disruption could have led to the formation of new species, and not surprisingly, different subspecies or, indeed, species of birds do sometimes occur on mountaintops that are now geographically isolated from one another. In other cases, differentiated taxa are found in close geographical contact; careful study of these contact zones has revealed hybridization in some instances and slight range overlaps without hybridization in others.

We have here a model for speciation, and hence faunal buildup, from within the system. The bearded helmetcrest *(Oxypogon guerinii)*, for example, is a small hummingbird inhabiting the high altitudes of four isolated Andean regions: three in Colombia (in the Santa Marta Mountains, the eastern Andes, and the central Andes) and one in Venezuela. Because the birds in each of the four allopatric, or geographically isolated, populations are morphologically differentiated, ornithologists once treated them as species; they are now considered subspecies. A possible sequence of events leading to this pattern is not hard to imagine. During a glacial period, a formerly widespread ancestor may have lived all along the central and eastern Andes and extended north into Venezuela and the Santa Marta Mountains. At this time, the paramo zone, in which the ancestral helmetcrest lived, was lower and more continuous than today. As glaciers melted, the paramo zone moved upward and became restricted to small pockets on mountaintops. Today, only the tallest summits, possessing the appropriate type of scrub and woodland, harbor the helmetcrest, whose isolation into four patches might have begun as recently as 20,000 years ago.

Should the climate become colder and glaciers push down the paramo habitats once again, I predict that the four isolates would come in secondary contact as their preferred habitat became continuous or nearly so. If complete speciation has taken place by the time of this secondary contact, geographical overlap might occur, and a region formerly inhabited by one ancestral species could be occupied by two daughter species. Such speciation may be a corollary of fluctuations in the environment that, in turn, resulted from climatic changes during the glacial–interglacial cycles.

Much of the previous reasoning is based on the assumption that a species' habitat requirements are largely invariant in time. This is not, however, necessarily the case. The bar-winged cinclodes *(Cinclodes fuscus)*, a brown, dipperlike bird that runs along the ground in search of insects and other small animals, is one of several wide-ranging high Andean birds that have geographically variable requirements. In some parts of the high Andes, this species of cinclodes is found in relatively dense and moist grassland, whereas elsewhere it inhabits much drier and more open scrub. If its habitat requirements vary geographically, there is no reason to believe they did not also vary temporally.

Another complicating factor is that species do not live alone; they share their preferred habitat with other species. To what extent are the habitat preferences of one species modified by the presence (or the absence) of another species, especially an ecologically similar relative? Does interspecific competition somehow rule how many species can live together in the same patch of habitat? While competition between species is an important ecological—and hence also evolutionary—factor, the precise role of competition is determining species ranges and influencing speciation (or extinction) is currently under great debate.

After an extensive analysis involving seventy-nine species of land and fresh-water birds endemic, or restricted, to the high Andes (representing about 44 percent of the total avifauna), Daniel Simberloff of Florida State University and I found good evidence that competition had been a causal factor in the distribution pattern for no more than six of the seventy-nine species. For example, in the case of canasteros (small, drab birds of the genus *Asthenes* that look like long-tailed wrens and build dome–shaped, covered nests), the distribution pattern is a mosaic in which several species do seem to replace each other at given study sites, rather than coexist. We concluded that the species of canasteros are mutually exclusive, a situation likely to result from interspecific competition. But in many other species, the primary causal agent of their Andean distribution may be associated with the glacial cycles. Competitive interactions may occur at all times, but in most cases, the ever-changing nature of the external environment would probably prevent interspecific competition from lasting long enough to settle distribution patterns as it has with the canasteros.

So far I have examined some mechanisms that help explain species dynamics in the high Andes once ancestral species occupied the newly opened habitats that formed after the mountains reached great heights. But how old is the fauna, and where did the ancestral stocks come from? A classic way of answering these sorts of questions has been, first, to look at levels of endemism—which can provide an indication of the age of the fauna and the relative importance of immigration and local speciation—and then to compare the fauna with others nearby that might have received or contributed species during the time being considered. What do these two aspects of faunal analysis tell us about high Andean birds?

In the high Andes, endemic taxa would be those restricted to above the timberline. The problem with this definition is that while some places have a sharp timberline, which effectively delimits two drastically different habitats (for example, cloud forest below, paramo grassland above, as in Colombia), in other areas (including the western slopes of the Peruvian Andes) there is no timberline. In the analysis that Simberloff and I carried out, we defined as endemic a taxon that lived only in the paramo and *puna* types of vegetation. We studied 147 species (the total avifauna includes about 180) at forty sites and found 79 to be endemic. In other words, nearly half the high Andean species are endemic, a rather high percentage.

In contrast, very few genera—4 out of 84—are endemic according to the same definition: a wader *(Phegornis)*, two hummingbirds *(Oreotrochilus* and *Oxypogon)*, and a finch *(Idiopsar)*. Finally, not a single avian family is endemic.

The low level of endemism at the genus level suggests a relatively young fauna. The high percentage of endemism at the species level, however, could indicate a rather old fauna. This apparent contradiction might be resolved if the species endemism is the result of a high rate of recent speciation. This speciation could be either local—that is, occurring only within the high Andes—or the result of immigration followed by differentation.

In birds, speciation is generally thought to be allopatric, involving the geographical division of a single species into two or more daughter species. Two good examples of local speciation in the high Andes include the white-throated and red-backed sierra finches, *Phrygilus erythronosus* and *P. dorsalis,* which look a little like oversized juncos, and the plain-capped and cinereous ground tyrants *(Muscisaxicola alpina* and *M. cinerea)*, slender, long-legged, terrestrial flycatchers. In both cases, the two species replace each other geographically except in a very narrow contact zone. A later stage in speciation is sympatry, when two formerly isolated species live together in the same habitat or share the same distributional range. I consider the plumbeous *(P. unicolor)* and red-backed sierra finches to be such a sympatric pair, as are the cinerous and *puna (M. juninensis)* ground tyrants.

To determine how much speciation has occurred *in situ,* we attempted to arrive at an estimate of the number of allopatric and sympatric species among the 147 birds in our study. Allopatric species representing recently completed speciation number about 12 to 20; sympatric species that may have been the result of local speciation number between 21 and 55. The variability in these figures reflects uncertainties in interpreting the data. A conservative estimate would be that about 20 to 25 percent of high Andean species are the result of local speciation. The rest, the overwhelming majority, therefore appear *not* to be the result of local speciation (unless their speciation is so ancient that it cannot be detected in present-day distribution patterns). This breakdown of species suggests, then, a high influx of immigrants, which in time have become endemic to the high Andes. In either case—local speciation or immigration—the age of the fauna remains largely unknown.

Determining where ancestral stocks came from can be equally complicated. We can compare lists of species living in paramo or *puna* vegetation with species living in other types of Andean vegetation (such as cloud forests and the open woodlands of the intermontane arid basins) or in the Patagonian steppes. This comparison reveals that the high Andes share more species and genera of birds with the Patagonian steppes than with any other ecological unit. This is not surprising since these two habitats are structurally very similar. The Andean *puna* and the Patagonian steppes, for example, share 48 percent of the total species found in them, while only 13 percent are shared by the *puna* and the geographically closer cloud forest. This would appear to vindicate F. M. Chapman's view, first published in 1917, that paramo-and *puna*-dwelling birds originated in southern South America, including Patagonia, and moved northward and altitudinally up-ward along the Andean corridor, colonizing as far as Venezuela and extreme northern Colombia. But Andean birds may also have contributed to the Patagonian fauna, and there may have been other routes of exchange as well. Birds of the *puna* and birds of the intermontane arid basins of the central Andes are somewhat related faunally; these regions share up to 30 percent of the species living in them, so that some exchange appears possible. Several high Andean genera and some species also occur in North America, which suggests another possible avenue of faunal flow. Examples among the species include the marsh hawk, the great horned owl, and the eastern meadowlark; among the genera, the stiff-tailed ducks, avocets, and flickers.

What can one conclude? Reconstructing the avifaunal history is clearly not easy. Readers interested in cautious scientific interpretations can turn to my technical papers; here let me indulge in speculation on a few of the birds found in the high Andes.

Once upon a time in the late Tertiary, about two to three million years ago, the Andes were not very tall mountains. In what is now Colombia and Ecuador to the north, their tops were either clad with cloud forest or, on the highest summits, covered with scrub. The high-altitude avifauna there included flycatchers, hummingbirds, tanagers, cotingas, and antbirds. Farther south, in Bolivia, plateaus were covered with savannas and woodlands, where caracaras, finches, and black-birds lived.

As orogeny proceeded and the mountains increased in altitude, local timberlines developed, and above them appeared a cold, treeless habitat dominated by grasslands and moorlands. The formation of this habitat, the precursor of today's paramo and *puna*, was the turning point in the faunal history of the high Andes. At first, few birds other than wide-ranging species that normally inhabit open spaces—ducks, coots, ibises, and waders in moist areas and lagoons, hawks in dry places and cliffs—could live in these habitats.

Some of these birds are good fliers and could have come from almost anywhere: from lower down the mountainsides, provided the slopes were not forested; from southern South America; and even from North America. Today's paramo and *puna* are inhabited both by relics of these early colonizations and by wide-ranging species that have invaded the high Andes more recently. Single stray birds from the lowlands, such as the wattled jacana *(Jacana jacana)* I saw at 10,000 feet in northwestern Argentina, are unlikely candidates for colonization. But many birds travel in flocks and can thus potentially stay and breed in areas far from their normal range.

Seed-eating finches and doves invaded the high Andes above timberline soon after seed-rich grassland habitats opened up. Little adaptive modification was required of these birds. Siskins—finches whose flocking behavior and wandering tendencies make them excellent colonists—could have come from almost any-where. Since they eat buds and other vegetable matter as well as seeds, they might easily have shifted from a forest to a shrub environment. In the *puna*, the endemic thick-billed siskin *(Spinus crassirostris)* appears to be restricted to some very high-altitude woodlands found within the grassland and scrub and may have derived from the smaller-billed hooded siskin *(S. magellanicus)*, a species with a wide range and broad habitat tolerance. The buntings, which include the yellow finches, may have moved in from open areas farther south, perhaps the grassy pampas or the Patagonian steppes, rather than the forested slopes below the newly formed paramo and *puna* zones. The arid woodlands of the montane basin, however, are almost as likely an original habitat for these stout-billed finches. As for the doves, they probably came from lower altitudes, and originally perhaps from North and Central America.

What about the small, brown ground-dwelling miners, other members of the diverse family Furnariidae, and the ground tyrants of the genus *Muscisaxicola?* These birds subsist on a diet that consists primarily of insects. Since none of the high Andean furnariids appear related to Central American stocks, they would seem to have come from farther south or southeast in South America, either the campos (grasslands) of Brazil, the pampas, or the Patagonian steppes. But they could also, in part, have evolved locally in the Central Andes, from stocks living in cloud forests down the slopes. The landscape probably progressed slowly enough from savanna or woodland to arid grassland or shrubland for such an ecological and evolutionary transition to be possible.

The *Muscisaxicola* ground tyrants are an extraordinary group of birds with unusual habits—for flycatchers. They live away from trees and hop and run along the ground on their long, thin legs. The proto-*Muscisaxicola*, having "discovered" the ground niche, then diversified into a large number of species by allopatric speciation, a good example of adaptive radiation. Several of the species are exclusively high Andean. Did they originally come from stocks already living in open spaces elsewhere, such as Patagonia? Perhaps, but not necessarily. While eight or so species of ground tyrants do live in open habitats along the Andean axis, from mountaintops in Colombia to southernmost South America, one species lives along rivers at the forested base of the Andes. The possibility thus exists that the ground tyrants evolved from lowland ancestors living in openings in forested habitats. The ecological preferences of the closely related chat tyrants (genus *Ochthoeca*) may shed light on the origin of the high Andean ground tyrants. Chat tyrants encompass all ecological intermediates, from cloud-forest species to arid-scrub species living high up in the *puna,* without much morphological or behavioral change. One species, the brown-backed chat tyrant *Ochtoeca fumicolor),* even shows intraspecific variation in habitat preferences: in the Venezuelan Andes, for instance, I have observed these birds in dense montane woodlands at about 11,000 feet and in increasingly more open habitats all the way up to 14,500 feet in desert paramo.

If the chat tyrants have invaded the geologically young paramo and *puna* habitats from lower-altitude montane forests, their distribution pattern may represent an earlier stage in the origin of ground tyrants. When it inhabited montane forests, the proto-*Muscisaxicola* of the high Andes might have looked and behaved like a chat tyrant, and then become modified during further evolution in the treeless *puna* and paramos. Eventually only the high-altitude species survived.

Figuring out what direction the shift in habitat preference took, however, is not as straightforward as it might seem. At face value, if there are currently more species in one habitat (*puna* or paramo) than the other (montane forest), the shift would appear to be from *puna* or paramo to forest. But speciation and extinction cycles, mediated by the external environment and modified by competition, could have turned things around so that today's lowland, forest-dwelling little ground tyrant could be either an ecological relict of the early forest history of the genus or a recent colonist from high-altitude, open habitats. Because it is almost as highly modified morphologically as the ground tyrants of high altitudes (or high latitudes, in Patagonia), I propose that the little ground tyrant is more likely to be a recent colonist.

This rough sketch gives an idea of the many different ways in which the birds of the high Andes may have evolved. Some species originated right there from ancestors that lived in more forested environments when the mountaintops were not so high; some are relative newcomers whose ancestors came from North America, from southern South America, or from the dry lowlands at the base of the Andes. Complicating the picture, some species are little different in morphology and behavior from closely related birds found elsewhere, while others have had a longer, more complex history. Challenged to reconstruct this history from often woefully inadequate evidence—involving shifts in habitat preference that trail far back in time, speciation between isolated mountaintops, and morphological and behavioral adaptations to the high Andes—the biogeographer may occasionally have cause to envy detective Hercule Poirot.

14

Antarctic Fossils and the Reconstruction of Gondwanaland

EDWIN H. COLBERT

At the present time geology is experiencing a revolution as profound as the one that shook biology a century ago, when Charles Darwin and Alfred Russel Wallace propounded the theory of evolution. This geologic revolution has to do with the theory of continental drift, which postulates that the continents have been mobile throughout the immensity of geologic time, rather than the stable elements they were so long thought to be. This is a revolutionary idea indeed, as the theory of organic evolution was a revolutionary idea. And as the theory of the evolution of life through natural selection gave man a new view of nature and of his place therein, so the theory of drifting continents has given man a new view of the earth on which he lives.

The idea of the evolution of life had been "in the air" for some decades before Darwin's *Origin of Species* was published in 1859. Likewise, the idea of drifting continents has been in the air for several decades—since the early years of this century and, in some respects, even before that. Darwin and Wallace independently gave initial form to the theory of evolution, but it was largely through the detailed and massive work of Darwin that the theory became established. Frank Taylor, an American, and Alfred Wegener, a German, independently gave initial form to the theory of continental drift, in 1910 and 1912, but it was largely through the efforts of Wegener and his brilliant follower Alexander du Toit of South Africa that the theory was developed in considerable detail.

For years, however, many, perhaps the majority of geologists throughout the world strongly opposed the theory of continental drift. Wegener and du Toit were ahead of their time; they had the concept, but they lacked the hard facts to give it a convincing basis. Now, within the past decade or so, facts have come to light in varied disciplines that have made continental drift not only a viable, convincing theory, but an exciting one as well. Continental drift is gaining ever wider acceptance among geologists the world around, and the modern geologic revolution is succeeding in a dramatic way.

To be valid, a theory must explain more or less satisfactorily all aspects of the phenomena with which it is concerned. For many years, numerous paleontologists—the students of ancient life on the earth—were not impressed by the theory of continental drift because they did not need it to explain the distributions of fossils on the continents. This was particularly true for the fossils of land-living vertebrates, backboned animals that moved from one place to another by dry-land routes. Paleontologists could explain the distributions of such animals through geologic time by postulating intercontinental movements across existing land bridges or across those that existed in the relatively recent geologic past: namely, the Panamanian Isthmus between the two Americas; the Bering bridge (presently interrupted by the relatively narrow and shallow Bering Strait) between the Eastern and Western Hemispheres; and of course the connections between Africa and the lands to the north. Australia, an island continent, was supposed to have had former connections with Asia. New Zealand and Madagascar, large islands near continents, were supposedly colonized by land-living vertebrates that adventitiously drifted to these isolated regions on masses of floating vegetation or logs. Such routes and means explained the distributions of ancient and present-day amphibians, reptiles, and mammals on the land masses of the earth.

Students of land-living vertebrates, however, largely ignored one continent— the island continent of Antarctica. It is true that today the edges of Antarctica are populated by such vertebrate animals as seals and penguins, as well as a few other birds, but the presence of these denizens of ocean and shore is readily explained. Aside from such marginal inhabitants, the absence of any true land-living vertebrates, recent or extinct, on the antarctic continent placed this great land mass, half again as large as the continental United States, generally outside the calculations of most students concerned with the distributions of ancient and recent tetrapods—the four-footed amphibians, reptiles, and mammals.

Then, in December 1967, Peter J. Barrett, a New Zealand geologist working in the Transantarctic Mountains about 400 miles from the South Pole, discovered a small fragment of a fossil lower jaw on the slopes of Graphite Peak in rocks of early Triassic age.

The specimen was too incomplete for close identification, but there could be no doubt as to its general nature: it was a portion of the lower jaw of a labyrinthodont amphibian, one of the tetrapods that lived during late Paleozoic and early Mesozoic times, from about 350 million to 200 million years ago. Here was a fossil of great significance, and it immediately drew attention from paleontologists, geologists, and biologists, as well as from the general public. Here was some slight indication that in the distant past Antarctica had been inhabited by land-living vertebrates.

Immediately, questions were raised. Was it not possible that the owner of this piece of fossil jaw had reached Antarctica by swimming across the surrounding ocean? Modern amphibians cannot tolerate salt water; if we apply the same physiological standards to the extinct amphibians, they could not have swum to Antarctica. But perhaps the oceans were less salty 200 million years ago. Moreover, some early Triassic amphibians have been found in marine sediments in Spitzbergen, although whether these fossils represent animals that habitually lived in the sea is open to question. At any rate, the evidence of one small jaw fragment, although most significant, was somewhat equivocal. More evidence was needed.

So it was that in October 1969, a group of us (William J. Breed of the Museum of Northern Arizona, James A. Jensen of Brigham Young University, Jon S. Powell of the University of Arizona, and myself) found ourselves at McMurdo Station in Antarctica, preparing to search for fossil vertebrates. We were part of a larger group of about twenty geologists and paleontologists, working under David H. Elliot, a geologist of note and a veteran antarctic explorer.

Our expedition was a gamble, and a costly one at that. We had no assurance that we would find fossils, and our chances for success seemed to diminish every day at McMurdo, as we waited through the weeks for storms to abate. It was the stormiest antarctic spring in years. Each day, as the winds howled past our huts, driving clouds of snow across the great ice shelf and Ross Island, on which the base is located, our long-laid plans for a concerted fossil hunt became increasingly tenuous and dislocated.

At last, however, on November 22, we flew into our camp near Coalsack Bluff, a nunatak (the exposed top of an isolated mountain largely buried in ice) on the edge of an ice field some 30 miles west of the mighty Beardmore Glacier, and some 400 miles from the South Pole. Elliot had chosen this locality primarily because it was a good spot for supply planes to land. We were to have helicopter support, and we proposed, first of all, to fly across the Beardmore Glacier (itself some 30 miles in width) to Graphite Peak, to begin our search where Barrett had made his discovery. We wanted to begin at a place where we knew a fossil had been found.

At this point serendipity took over. Coalsack Bluff was about five miles away across the ice, and the helicopters had not as yet arrived, so on the first day in camp some of our group went over to Coalsack Bluff because it was there. Almost immediately we found fossil bones in some low cliffs, exposed on the far side of the nunatak. Before the day was out, nearly thirty fossils had been located along a half mile or more of cliff exposures. From that day until the end of our stay, we spent most of our time excavating fossils from the sandstone cliffs of Coalsack Bluff.

Something should be said about the locale at which we were excavating the fossil bones. A frequent question asked of us on our return was: "How did you find the fossils? Did you dig down through the ice for them?"

Antarctica is commonly pictured as a great, ice-covered continent, and so it is over much of its extent. But in the Transantarctic Mountains there are extensive cliff exposures where the high mountains rise above the level of the glaciers and ice fields. The mountains in large aspect form a continuous range across the continent, but at many places there are outlying nunataks, and Coalsack Bluff is one of these. The north side of Coalsack Bluff is a long slope, largely free of snow and ice. Its lower portion is composed of dark shales with layers of coal belonging to the Permian Buckley Formation, containing, in places, abundant fossil leaves of the characteristic Gondwana plant *Glossopteris*. Above these shales and coals is the Fremouw Formation of early Triassic age, an alternation of sandstones and shales. The sandstones, generally brown or gray in color, stand up as low cliffs, and the shales form the slopes between them. There are three such sandstone cliffs, one above the other, on the slopes of Coalsack Bluff. Finally, capping the nunatak and appearing on its slopes as intrusions, are thick volcanic rocks. These dense rocks, broken and weathered into highly polished slabs, cover much of the slope of Coalsack Bluff.

The weathering processes in Antarctica are unlike those in other parts of the world. Central Antarctica is a desert, with an amazingly scanty annual increment of moisture. Temperatures are low, so there is little thawing. Much of the erosion in the Transantarctic Mountains is effected by wind, wind that sweeps off the polar plateau in fierce gales, driving the dry snow in horizontal clouds. These are the ground blizzards of Antarctica. At extremely low temperatures the snow is so hard and dry that it acts very much like wind-driven sand. The force of these blizzards polishes the hard volcanic rocks and cuts them into weird shapes. On cliffs and exposed slopes, such as the long slope of Coalsack Bluff, the winds clear the snow away, leaving the rocks exposed—a fortunate circumstance for the fossil hunter. (In another sense the antarctic winds were anything but fortunate for us; they were our worst enemy in the field, making our work frequently difficult and at times impossible.)

The sandstone cliffs from which we collected the bones were the solidified remains of ancient stream channels. We were dealing with sediments laid down in streams, sediments containing the bones of amphibians and reptiles that had lived in and along the edges of the streams.

It soon became evident that we were finding the bones of labyrinthodont amphibians and mammal-like reptiles. On December 4, a portion of a skull was discovered that proved to belong to the reptilian genus *Lystrosaurus*. The discovery of *Lystrosaurus* with other mammal-like reptiles and with labyrinthodont amphibians indicated that we had found in the Transantarctic Mountains an association of amphibians and reptiles similar to that occurring in the Lower Triassic beds of South Africa, designated the *Lystrosaurus* fauna. The *Lystrosaurus* fauna has also been found in the Lower Triassic sediments of peninsular India, and in Sinkiang and Shansi, China.

We never did go to Graphite Peak, in part because we were completely busy at Coalsack Bluff, and in part because of helicopter troubles. Nor did we go, as originally planned, to McGregor Glacier, some 150 miles southwest of Beardmore Glacier, partly because of problems of logistical support, and partly because of the delays resulting from the bad weather that had plagued us. McGregor Glacier was reserved for the following season.

The next season came, the austral summer of 1970–71, and with it the campaign at McGregor Glacier. This time the fossil hunters were led by James W. Kitching of the Bernard Price Institute of Paleontology, Witwatersrand University, Johannesburg, assisted by John Ruben of the University of California and, for a short time, by Thomas Rich of Columbia University. Again David Elliot led the entire party working at McGregor Glacier. Kitching was the best possible man to continue the search for fossils in the Transantarctic Mountains. He has spent a lifetime working in the Permian and Triassic sediments of South Africa, and it is fair to say that no other paleontologist alive can equal his experience in the search for the Permo-Triassic amphibians and reptiles that occur so abundantly in the African Karroo sequence. We knew by then that the fossil tetrapods of Antarctica are of close African relationships; Kitching was the logical man to look for additional and perhaps more complete *Lystrosaurus* fauna fossils in the Fremouw Formation.

History repeated itself. On the first day in camp at McGregor Glacier, James Collinson, a geologist, discovered in the rock a skeletal imprint of *Thrinaxodon*, a mammal-like reptile associated with *Lystrosaurus* in the African sediments. From then on fossils were continually found in the McGregor Glacier region, many of them articulated skeletons or partial skeletons. The fossils found at McGregor Glacier show that in addition to *Thrinaxodon*, there were in Antarctica other mammal-like reptiles similar to those found in the *Lystrosaurus* fauna of South Africa, and also such *Lystrosaurus* fauna tetrapods as the little reptile *Procolophon;* small eosuchian reptiles more or less ancestral to lizards; various thecodont reptiles especially characteristic of Triassic sediments; and labyrinthodont amphibians that may be compared not only with the African *Lystrosaurus* fauna amphibians but also with Lower Triassic amphibians found in Australia. Consequently, it is now apparent that there was a fully developed *Lystrosaurus* fauna living in Antarctica in early Triassic time, a fact of particular significance.

In the first place, the presence of a diversified *Lystrosaurus* fauna in Antarctica indicates beyond any reasonable doubt that there was a dry-land connection between the present south polar continent and southern Africa. *Lystrosaurus, Thrinaxodon,* and the other mammal-like reptiles that have been found in Antarctica, as well as *Procolophon,* the thecodont and eosuchian reptiles, and the amphibians could have moved back and forth between what are now the Transantarctic Mountains and the Karroo Basin only across a land route. In the second place, the broad spectrum of the *Lystrosaurus* fauna in Antarctica is almost certainly an indication of a wide dry-land avenue, allowing the entire fauna to spread from Africa to Antarctica (or vice versa). Such a complete representation of the fauna in both areas is strong evidence against a narrow isthmian bridge connecting ancient Antarctica with ancient Africa, for we know from modern examples (from the Panamanian Isthmus, for example) that an elongated, narrow bridge acts as a zoological filter, permitting some animals to migrate along its length but excluding other animals from using it. No such filter effect is apparent in comparing the *Lystrosaurus* fauna fossils from the Transantarctic Mountains with those from South Africa. Indeed, the close resemblances between fossils in the two regions, extending down to a similarity of species among various genera, is evidence that in early Triassic time Antarctica and southern Africa were probably integral parts of a single continental land mass. The presence of the *Lystrosaurus* fauna in these two regions is probably a manifestation of a single fauna within the limits of its natural range. Again, the *Lystrosaurus* fauna, composed of various reptiles, some of them of considerable size, and of large amphibians as well, is obviously an assemblage of tropical or subtropical animals. This means that Triassic southern Africa and Antarctica probably were in latitudes lower than those they now occupy.

This brings us to the subject of Gondwanaland. The name was coined in the latter part of the nineteenth century by the Austrian geologist Eduard Suess to designate a hypothetical gigantic continent, embracing the modern continents of the Southern Hemisphere, and extending across the equator to include the peninsular portion of India as well. This great ancient continent was considered by many geologists as useful, and perhaps necessary, to explain many similarities among the rocks and fossils of the Southern Hemisphere continents and of peninsular India. Some of the early believers in Gondwanaland pictured it as an immense east-to-west land mass, including India and the Southern Hemisphere continents as they are now placed. Gondwanaland, they thought, subsequently disappeared as an entity by the foundering of large portions of land into the oceans, leaving the present continents as isolated remnants. Other students, who found it difficult to visualize the sinking of such great expanses of land beneath the ocean, thought of Gondwanaland as being composed of the present southern continents and India as we know them, all connected by long and relatively narrow land bridges. Then Wegener, and after him du Toit, introduced a new concept, namely that the several continents making up ancient Gondwanaland were at one time contiguous, and that subsequently the ancestral land mass fragmented, its component parts drifting to their present positions. (A similar parent continent, Laurasia, has been proposed for the Northern Hemisphere, its subsequent fragmentation and the drift of the fragments having produced North America, Greenland, and most of Eurasia.)

Early opposition to the theory of continental drift included, of course, opposition to a Gondwanaland formed by the modern Southern Hemisphere continents and peninsular India. But modern geologic findings strongly support such an ancient continent, its eventual fragmentation, and the drift of its fragments to their present positions. The many facts that point to this sequence of geologic events are too complex and involved for elucidation here. Suffice it to say that the complementary theories of plate tectonics and of sea-floor spreading, which stipulate that the crust of the earth is composed of a number of gigantic plates that are constantly in motion, provide the mechanism, previously lacking, to explain continental drift.

Our present concern is how the fossil evidence accords with the concept of Gondwanaland and the theory of continental drift. Do the distributions of early land-living vertebrates, especially those that have been found in Antarctica, support Gondwanaland and drift?

As we have seen, the fully developed presence of the *Lystrosaurus* fauna in Antarctica indicates that Antarctica and southern Africa were joined along a broad front. The same is true to a somewhat lesser degree for peninsular India, where the *Lystrosaurus* fauna is found partially represented. If present-day Africa, Antarctica, and peninsular India are joined according to the similarities of their outlines at a depth of 1000 fathoms, the "fits" between them are remarkable. This is particularly true for the edge of the African continent between Durban and

Mozambique and for Antarctica along the Weddell Sea and the Princess Martha coast. Such a fit affords a broad connection between the two land masses, making of them essentially a single land.

And such a fit, together with the fit of peninsular India between Antarctica and eastern Africa, brings the localities of the *Lystrosaurus* fauna in these now widely separated continents all within about 2000 miles, or less, of each other. This distribution suggests a very reasonable range for a terrestrial vertebrate fauna, as judged by modern standards. A single species of *Lystrosaurus* is present in Antarctica, Africa, and India; it seems quite probable that this species on the modern continents represents the disruption of what was once a relatively compact range of distribution. (The presence of elements on the *Lystrosaurus* fauna in China has yet to be explained, but at the moment it would appear that facts are accumulating that will account very satisfactorily for the Chinese occurrences of these early Triassic tetrapods.)

So it is that the discovery of the Lower Triassic *Lystrosaurus* fauna in the Transantarctic Mountains is a paleontological development of prime importance. It helps prove the close connection of Antarctica and southern Africa in Triassic times. From this demonstration of faunal and continental relationships one proceeds to the conclusions that there was such an entity as Gondwanaland, that Gondwanaland was broken asunder, that its fragments drifted apart, and that Antarctica, once the habitat of tropical or subtropical amphibians and reptiles (and abundant plants as well), came to occupy a position in a climate quite inimical to the life that had once flourished in benign temperatures. Other geologic and paleontological facts support the conclusions drawn from continental outlines and the occurrences of the *Lystrosaurus* fauna, such as the general expression of Permian and Triassic geology in southern Africa and Antarctica, the presence of extensive volcanic rocks in the two continents, and the development of fossil plants in these areas. But the occurrences of the *Lystrosaurus* fauna are also important; they give solid evidence for land connections. The evidence of geophysics involves certain assumptions, as does that of geology. The evidence of the fossil plants is strong, but there is always the possibility (although according to the paleobotanists, a very slim one) that these plants may have been distributed in part by wind-borne transportation of seeds. The evidence of the land-living tetrapods, present in the two regions as fully developed faunas, cannot be denied. These animal assemblages most surely had to move from the one region to the other on dry land.

The recognition of a *Lystrosaurus* fauna in the Transantarctic Mountains is of significance not only because it adds a large dimension to our knowledge of ancient life on what is now the South Polar continent but also, as we have seen, because of the strong confirmation it lends to continental drift and to the former existence of Gondwanaland. Important as the discoveries of the past two years are, however, they have merely scratched the surface of antarctic paleontological riches. For riches there are, in the form of numerous untouched exposures of the Fremouw Formation containing abundant fossils.

Much progress has been made in the elucidation of ancient life on Antarctica since those tragic days sixty years ago, when Scott and his companions unsuccessfully struggled back from the South Pole, dragging their sledge loaded with survival gear and with some 25 pounds of precious fossil plants. Even more progress lies ahead. In Antarctica surely are paleontological answers to many questions regarding the evolution and distribution of life on the ancient continent of Gondwanaland.

15

An Extravagance of Species

NILES ELDREDGE

In the summer of 1972, a colleague asked me to help with a routine identification of a fossil owned by LeGrand Smith, an American missionary living in Bolivia, who was passing through New York en route to Czechoslovakia with a load of fossils. The single specimen he brought us resembled nothing so much as a rounded lump of hardened clay, and we first pronounced it "inorganic." But a thin ridge with what looked like the lenses of the eye of a trilobite (an extinct relative of crabs and shrimp) finally convinced us that there must be something organic hidden inside.

A month of tedious, painstaking preparation with a sharp needle eventually uncovered a fossil, although not a trilobite. It was so odd that we couldn't tell exactly what kind of fossil it was. After some detective work in the literature we finally established that it was an exquisite, uncrushed specimen of a rare proto-horseshoe crab. Only three highly distorted and totally flattened specimens even remotely like the Bolivian creature had ever been seen before. The crushed specimens came from Lower Devonian slates of southern Germany. Because of the excellent preservation of the Bolivian material, anatomical details of this group of early horseshoe crabs were clarified for the first time, which enabled us to reassess the early stages of evolution of horseshoe crabs.

But the really appetizing part of the encounter with Smith was his offhand assertion that he had thousands of Devonian fossils in his collection. His collaborator, Dr. Leonardo Branisa, then of the Bolivian Geological Survey, had amassed a still larger collection and had recently sold several thousand specimens to the Smithsonian Institution. No paleontologist even remotely interested in the fossils of a region could ignore or resist the promise of an open pipeline into what looked like enormous fossil wealth. Smith, Branisa, and I joined forces, and I soon found myself inundated with Bolivian trilobites.

Nothing delights a paleontologist more than confronting a rich collection of well-preserved fossils, and the Devonian rocks of the southern continents, the source of my trilobites, are famous for their fossils. They have been studied sporadically, mostly by northerners, for more than 100 years and until recently seemed fairly well known. Famous collecting sites in Bolivia, Brazil, and South Africa have long since been carefully picked over, and descriptions of their fossils have filled the pages of at least fifty articles and monographs. But now that remoter areas, particularly of the South American Andean Cordillera, are being frequented by fossil hunters, we find that we have only scratched the surface.

A flood of new fossils has easily doubled the number of known species of marine invertebrates. But beyond the simple pleasure of filling a large gap in our knowledge of the history of life, this new fossil material has raised a number of questions about the origin and history of a vast territory of the world and its inhabitants during a 40-million-year period of time that spans the Upper Silurian and Lower Devonian periods, from about 410 to 470 million years ago.

The area involved is the more southerly part of famed Gondwana, the once-unified landmass that drifted apart to become South America, Africa, Antarctica, India, and Australia. And the animals involved are shelled invertebrates—past denizens of shallow marine habitats. Dominant among these invertebrates are the trilobites.

Trilobites, including the ones from Bolivia that I studied, pose some fundamentally important evolutionary puzzles, puzzles whose solutions demand changes in evolutionary theory. It would be tempting to offer a scientific detective story of my work, with patterns in the raw data suddenly emerging to suggest a new theory. But no scientist ever accumulates facts blindly. There are always new ideas emerging that influence the way we look at new, and old, facts.

When I first became aware of the Devonian trilobites of the Southern Hemisphere, new ideas of the earth's history, such as the notion of continental drift, were in the breeze. These ideas were having their effect in other areas, particularly on explanations of how animal and plant distributions change through time (historical biogeography). I was entranced with a new and, in some respects, radically different way of thinking about the evolutionary process that questions the role of natural selection in creating new species.

Theoretical ferment on evolution in the Southern Hemisphere is not new. The history of research in the ancient geography and the fossil animals of the southern continents has been tortuous and filled with controversy. To begin with, the notion that Africa and South America were previously joined has only recently gained widespread acceptance. For years, just a few solitary voices supported this idea, including those of the South African Alexander du Toit and the American Kenneth Caster, both of whom were familiar with the similarity between the fossil faunas of South America and southern Africa. Others believed the continents had always been far apart.

Depending on which view of geographical history is adopted (stable continents always separated or Gondwana now fragmented), paleontologists naturally develop rather different ideas about the history of distributions of animals and plants. If the pattern to be explained is the presence of similar organisms on two forever widely separated continents, our explanation invokes long-distance dispersal over the barrier of the South Atlantic. For marine invertebrates, we must invoke current patterns capable of wafting larvae long distances. The invention of these current patterns sometimes stretches the principles of oceanography beyond endurance.

Dispersal need not be invoked if we let the continents do the moving. In this approach, major changes in the distributions of organisms over the globe are passive reflections of actual changes in the physical geography itself.

Thus we have two conflicting sets of explanations for historical changes in animal and plant distributions; both have come into play in the analysis of the history of the 380-million-year-old fossils from the Southern Hemisphere. Now there is a consensus that the similar fossil faunas scattered over the southern continents reflect fragmentation of Gondwana rather than long-range larval dispersal.

What is the evolutionary history of the Gondwana trilobites? Most, roughly thirty of fifty genera, belong to the Calmoniidae, a single family that occurs nowhere else and whose numbers offer an excellent chance to look for evolutionary patterns.

Apart from their dominance, the calmoniid trilobites are a fascinating lot. Derived from a garden variety, run-of-the-mill simple form, the sixty or so species known to date diversified into one of the more spectacular radiations of the entire 400-million-year history of trilobites. A few retained the general ancestral simple form. But most modified the old form, and some became, even by trilobite standards, downright bizarre. I have calmoniids with spines added to the front, or to the sides and middle of the head, or around the margins of the tail. I have others up to seven inches long—quite a length for a trilobite, most of which are less than one and a half inches long.

One group, which seems to be natural (meaning that all its members are closely related and form a single sublineage of the family), invented a unique anatomical plan. The genera in this group do not resemble any other kind of trilobite particularly closely. They are immediately identifiable as unique forms. *Deltacephalaspis*—the long name simply means "triangular-shaped head with spine"—is both typical of this group and highly unusual for trilobites in general. The immense spines sticking out from both sides of the head are particularly noteworthy. To guess with any certainty about how trilobites lived is difficult in itself, as their appendages, used for food gathering and locomotion, are almost never preserved. But I am totally at a loss to explain strange structures like the pair of enormous, hornlike spines on either side of the head of *Deltacephalaspis*.

The spines could have been stabilizing structures used in swimming or in righting the animal if it were ever tipped over. (Think of the difficulty a wingless insect has in getting back on its feet after falling on its back.) But with no clear living analogs, such speculation, however entertaining, is fruitless. For *Deltacephalaspis* and its relatives (such as the spine-snouted *Schizostylus* and *Cryphaeoides*), we must be content with just describing the array of different kinds, assuming that there were functional reasons underlying their evolutionary development.

The evolutionary pattern of diversification in the other major calmoniid group is more readily explained—at least in general terms. Also evolutionarily natural and coherent, this group shows, if anything, an even greater range of anatomical configurations than the first. But the kinds of anatomical modifications developed by members of this group are rather familiar. They are reminiscent of trilobites in other families. *Bouleia dagincourti*, for instance, is similar to *Phacops*, a trilobite that is also found in Gondwana.

I stumbled on this similarity before I became deeply involved with the Bolivian fossils. At the time, I was working on *Phacops*. Browsing through the trilobite collections at the American Museum of Natural History, I found a particularly beautiful specimen of *Bouleia*, which was said to be closely related to *Phacops*. The specimen was well enough preserved to show that its resemblance to *Phacops* was entirely superficial. The only special similarities between them were a large, bulbous head and a small, rounded tail. Here, apparently, was a case of evolutionary convergence, in which remotely related organisms evolve somewhat similar structures, adapting to similar environmental demands.

In this second group of calmoniids there are several examples of convergence. *Metacryphaeus*, a calmoniid, is very much like the dalmanitid trilobites common in Devonian faunas elsewhere in the world. Another case of convergence is the similarity of *Schizostylus*, with its long snout and cheek spines, to the rare *Probolops*, both calmoniids. The head of *Probolops* is modified into exactly the same form as *Schizostylus*. Since the two are found in the same beds and *Probolops* is extremely rare, it may well have gained some advantage by mimicking *Schizostylus*.

Such similarities offer some insight into the evolutionary history of the Gondwana trilobites. But we are left to explain the most striking fact about these fossils—their diversity. What does all this variety suggest? How and why did ancestral trilobites evolve into so many kinds of descendants, so anatomically varied?

Standard evolutionary theory focuses on anatomical change through time by picturing natural selection as the agent that preserves the best of the designs available for coping with the environment. This generation-by-generation process, working on small amounts of variation, is thought to change, slowly but inexorably, the genetic and anatomical makeup of a population.

If this theory were correct, then I should have found evidence of this smooth progression in the vast numbers of Bolivian fossil trilobites I studied. I should have found species gradually changing through time, with smoothly intermediate forms connecting descendant species to their ancestors.

Instead I found most of the various kinds, including some unique and advanced ones, present in the earliest-known fossil beds. Species persisted for long periods of time without change. When they were replaced by similar, related (presumably descendant) species, I saw no gradual change in the older species that would have allowed me to predict the anatomical features of its younger relative.

The story of anatomical change through time that I read in the Devonian trilobites of Gondwana is similar to the picture emerging elsewhere in the fossil record: long periods of little or no change, followed by the appearance of anatomically modified descendants, usually with no smoothly intergradational forms in evidence.

If the evidence conflicts with theoretical predictions, something must be wrong with the theory. But for years the apparent lack of progressive change within fossil species has been ignored or else the evidence—*not* the theory—has been attacked. Attempts to salvage evolutionary theory have been made by claiming that the pattern of stepwise change usually seen in fossils reflects a poor, spotty fossil record. Were the record sufficiently complete, goes the claim, we would see the expected pattern of gradational change. But there are too many examples of this pattern of stepwise change to ignore it any longer. It is time to reexamine evolutionary theory itself.

There is probably little wrong with the notion of natural selection as a means of modifying the genetics of a species through time, although it is difficult to put it to the test. But the predicted gradual accumulation of change within species is seldom (if ever) encountered in our practical experience with the fossil record.

The problem appears to be this: focusing attention purely on anatomical (and underlying genetic) change ignores a fundamental feature of nature—the existence of species. Species are reproductive communities. They are held together by a network of parental ancestry and descent and separated from other, similar networks of parentage. They are coherent entities in space and—and this is the crucial part—through time as well. Species have origins, histories, and extinctions. They may or may not give rise to one or more descendant species during the course of their own existence.

The internal cohesion of a species predicts the sort of lack of anatomical change during its history that we see so often in the fossil record. And speciation, the process of creating a new, descendant reproductive community from an ancestral one—in a way other than that proposed by natural selection—must be a crucial step in the evolutionary process.

There are three or four commonly accepted, slightly different notions on how speciation occurs. The most favored theory of speciation for active, complex animals such as trilobites is geographical, or allopatric, speciation—what nineteenth-century biologists called vicariance. The basic idea of this model is simple: the easiest way to achieve reproductive independence between an ancestral species and its descendant is to segregate a segment from the main part of the ancestral species. Speciation occurs if enough behavioral or anatomical differences develop to guarantee that the two species will not hybridize should they one day come to inhabit the same area.

Other speciation theories hypothesize ways in which a new reproductive community—a fledgling species—can bud off from its parent without being isolated geographically as an initial step. But none of the various theories see *natural selection* changing a species' adaptations enough to develop a "new" species from an old one. Rather, speciation is a largely accidental fragmentation of a once coherent reproductive community.

During such events, behavioral, anatomical, and genetic change (sometimes quite rapid) may occur. Such change must occur if we are to be able to tell the daughter species from the parent. And the process of natural selection must account for part of such change. But natural selection per se does not work to create new species. The patterns of change in so many examples in the fossil record is far more a reflection of the origin and differential survival (selective extinction) of species than the inexorable accumulation of minute changes within species through the agency of natural selection.

The evolutionary pattern of the calmoniid trilobites is decisive: an evolutionary theory that maintains the primacy of pure natural selection and entertains no other mode of change is incomplete. In fact, the Devonian trilobites of Gondwana support, on several levels, the importance of geographical differentiation and speciation emphasized so long ago by Darwin and some of his contemporaries.

First, speciation is the key to understanding the pattern of anatomical change actually seen in the fossil record. Since geographical isolation is the simplest way to bud off a descendant from an ancestral species, we look for adjacent regions harboring slightly different, closely related races and species. Within Bolivia itself, there is a pattern of geographical variation from north to south among many of the calmoniid species.

On a grosser level, three major subregions of the Devonian Gondwanan province can be recognized, each with its characteristic assemblage of species related to, but usually not the same as, those in the other subregions. Patterns of relationship among the species of calmoniids indicate that the Andean subregion may be more closely affiliated with southern Africa than either is with the stable continental interior subregion of South America. The Gondwanan map shows a possible explanation for this pattern: current patterns reconstructed on the basis of principles of physical oceanography indicate a more complete connection between the Andes and southern Africa than between either of those areas and continental South America.

Finally we return to the grand pattern we took up at the outset: the entire Gondwanan province is unique. It is an area that was so isolated from the rest of the world that it peopled its seas almost entirely with its own home-grown products. There is no clearer picture of the relation between geography and evolution than the pattern of the origin, history, and demise of the Devonian trilobites of ancient Gondwana.

16

The Elusive Eureka

NILES ELDREDGE

The tendency of scientists to make pronouncements from remote Olympian heights lends credence to the charge that science is just another belief system, no different from any other. The current efforts of fundamentalist Christians to inject the biblical story of Creation into biology curricula across the nation exploit this very impression. Creationist cries of "equal time" appeal to people's sense of fair play because science, like religion, often appears authoritarian. And it is true that science has become so complex that no one can hope to grasp it all. Most of us are content to believe that the earth is round, even though we are personally unable to make the necessary observations that verify it. But before a scientific idea can earn this kind of routine acceptance, it must be subjected to the scrutiny of the "scientific method."

The scientific method is not some mysterious form of higher cerebration. Its principle is simply that our ideas about the universe—how it is constructed, how it came to be—must conform to our observations. Scientists test an idea by predicting what they should find if it is true, and their understanding of the natural world is refined in a constant interplay of ideas and observations. This very human exercise is very different from the unquestioning acceptance of "revealed truth." A scientist is prepared to evaluate any new piece of evidence, even if it challenges an orthodox theory. Sometimes the theory will be discarded or modified; sometimes the evidence will be thrown out of court. But in either case something will have been gained. I have here a modest example of how the system works; perhaps it will give an idea of what scientists actually do for a living.

No one (presumably) would dare write an article for a science magazine without being reasonably sure of his or her ground. So when I reported in the July 1980 issue of *Natural History* on distribution patterns of Devonian trilobites in the Southern Hemisphere, I was relying on 150 years of accumulated paleontological research, as well as ten years of my own contemplation of these fossils. The picture I presented was a neat one. Trilobites are extinct relatives of crabs and shrimps; those that I wrote about lived in what was evidently a cold-water habitat during the Devonian period, 395 to 345 million years ago. Their fossils are most numerous in Andean South America from southern Peru south through Bolivia and Argentina. They also occur in less exuberant numbers in the Amazon Basin, portions of southern Brazil and Uruguay, the Falkland Islands, and South Africa. The most reasonable explanation for this distribution is that there once was a great, partly submerged landmass, which geologists call Gondwana, that broke up to form South America, Antarctica, Africa, India, and Australia. The widely scattered localities that yield these cold-water Devonian fossils were once much closer together, and all were south of 60° south latitude.

As the article for *Natural History* was in its final editorial phase, however, I received a letter from Alfonso Segura Paguaga, a Central American geologist living in Esparza, a rural area of Costa Rica near the Pacific coast city of Puntarenas. I was startled at his announcement of the discovery of a Devonian trilobite in Costa Rica. No rocks even remotely as old had ever been found there. Paguaga wanted me to examine the specimen, proposing to name it *Phacops esparsocostariccensis*. I hurriedly wrote back that I would be delighted to work on the trilobite, the possibility of a bona fide discovery lurking in the back of my mind.

Paguaga arrived in my office with the trilobite, as well as fossils from four separate outcrops in the environs of Esparza. Eagerly unwrapping the trilobite first, I was flabbergasted. There before me lay a typical example of *Metacryphaeus tuberculatus*—a very common trilobite, indeed, but one known only from southern Peru and Bolivia. "Were I not informed," I wrote rather formally in my report to Paguaga, "that the specimen came from Costa Rica, I would without hesitation say it came from the region of Padilla, Bolivia."

Two things were puzzling about the fossil. The first was its age: as far as I knew, Venezuela and Colombia in South America, and Mexico in North America, were the closest places to Costa Rica that had ever produced a Devonian trilobite. Second, the specimen was not the type of trilobite we would expect to find even in those areas. Trilobites from northern South America are of the same basic stripe as our Northern Hemisphere Appalachian trilobites—and radically different from those, such as *Metacryphaeus*, that lived in the cold seas that bathed Gondwana.

The geologic history of Central America and the Caribbean is extremely complex and as yet only partly unraveled. What we seemed to have was an "island" of Devonian Gondwana—a portion of Andean South America that had somehow become detached and incorporated as an exotic chunk into what later became Costa Rica. A radically new view of shifting fragments of continental plates colliding to form portions of Central America started to take shape in my mind. It was too late to stop the *Natural History* presses, but I began to envision a quick report to *Science* or some other prestigious "what's the latest development in science" journal.

But there was a small fly in the ointment. Paguaga said he got the trilobite from a small boy who had brought it to school. The boy had found it on a local roadside. If we were to tell the world of an exotic Devonian "island," we had to be able to point to a mass of rock and say, "There it is." Another specimen Paguaga brought with him seemed promising: a chunk of brown siltstone about the size of a silver dollar. It had shell fragments that looked the right age on one side, and a gorgeous fossil, which I took to be a previously unrecorded variety of phyllocarid, a type of crustacean especially well known in Devonian rocks. About an inch long, the symmetrical structure was elaborately sculptured with a variety of bumps and furrows reminiscent of, although a bit more complicated than, the carapaces of known phyllocarids. We seemed to be in good shape, but Paguaga understood the necessity for ironclad evidence that the trilobite was indeed a native of Esparza.

Shortly after his return home, Paguaga went to work, and I was soon bombarded with packages of Costa Rican fossils. Concentrating on the outcrop that had produced the crustacean, he found more than fifty complete fossils and shipped them all to New York. To my chagrin, all proved to be clams and snails no older than twenty or thirty million years—very much the sort of thing one was supposed to find in Costa Rica.

With my doubts deepening, I retrieved the "crustacean" from the collection of Costa Rican fossils, stared at it hard, and realized it was not a phyllocarid after all. It was (and I am no botanist) apparently some sort of nut or seed pod. Groaning, I realized that my enthusiasm had carried me away. I had fantasized a Tertiary fruit into a Devonian crustacean! All of a sudden I was quite glad I hadn't batted out a note to *Science* trumpeting our wonderful discovery.

The real story of the Esparza trilobite is not exactly fraught with scientific significance. It turned out that my first suspicion about the origin of the fossil was right. A U.S. citizen, leaving La Paz, Bolivia, to live in Costa Rica, bought the trilobite from an "old Indian" as a souvenir. (I hear that these trilobites have been for sale all over La Paz in recent years.) He had a minor car accident on a Costa Rican road one day, and the trilobite was lost—only to be found by the little boy. All the details were uncovered by Paguaga—a dogged truth seeker if ever there was one—who advertised in the papers and tracked down the American, still living in Costa Rica. Of the thousands of Andean fossils in my office, none has taken a more circuitous route from the mountains of Bolivia to our labs in New York.

The whole episode, as absurd as it may seem, really does show how science works. The misplaced trilobite challenged several longstanding notions I had confidently recounted in my article for *Natural History*. Either the ideas or the trilobites had to give.

In the end, I was sorry not to have been a party to a real discovery. It would have been nice to have had the chance to exclaim "Eureka!" the way scientists are popularly supposed to do. You know the image. After years of tedious measurement, a new generalization on the properties of matter suddenly dawns on the patient physicist; after much careful planning, the pith-helmeted coot in khaki shorts bags a "hitherto-unknown-to-science" butterfly in the depth of Amazonia. Eureka! But, alas, eurekas are hard to come by. The true stuff of science remains the sure and steady checking of ideas against worldly realities.

PART 5

LIVING FOSSILS

Evolution means "change" to most people. Why, then, in a book on evolution should we be concerned with "living fossils"—creatures that appear not to have changed from their ancestral state for periods of time that seem formidable even in geological terms?

The theory of evolution we have been piecing together so far has as its core the notion of adaptation through natural selection. Darwin succeeded in implanting the concept so well that we came to think of adaptive evolutionary change as inevitable, and not the impossibility nearly everyone believed it to be prior to the appearance of *On the Origin of Species*. But if change is literally built into the system, how then do we explain these anomalous cases of "arrested evolution," these living fossils who have resisted the irresistible?

Evolution is best thought of as maintenance, modification, and transmission of genetically based information. Modification, the transformation of organismic features and their underlying genes, is only part of the story. There are many reasons why organic features should not become modified, many of which are broached in the following essays. Understanding why evolutionary change does not occur, why things remain stable, is to take us a long way down the path toward understanding how and why change occurs—when it occurs at all.

"Living fossils" is a phrase that actually encompasses a variety of similar, yet slightly different, phenomena. Some biologists see the phrase as devoid of meaning, or at the very least misleading. And it is true that, if what is meant as a definition is that species tend to remain unchanged *as the very same species* for tens, or even hundreds of millions of years, "living fossils" is an empty concept indeed. But all that is usually meant by the phrase is that some creatures alive today are astonishingly reminiscent of relatives that lived truly long ages ago. Evolution seems to have passed them by, all the while greatly modifying their close collateral kin as time has gone by.

The first three essays set out the details of particular cases of living fossils. Howard L. Sanders recounts the history of discovery of the primitive crustacean *Hutchinsoniella macracantha*, found living virtually under our noses in Long Island

Sound. *Hutchinsoniella* actually violates one of the common aspects of a living fossil: it has, so far, not shown up in the fossil record! As Sanders discusses, it is actually *more* primitive than its most similar fossil relative. *Hutchinsoniella*, though, is immediately recognizable as the most primitive of all known crustaceans, fossil or recent, because of the extensive work already done on crustacean comparative anatomy and embryology. And this remarkable little animal has also helped us understand arthropod evolutionary history—phylogeny—more clearly; Sanders makes a particularly interesting argument that trilobites and crustaceans are actually more closely related than prevailing wisdom had acknowledged prior to the discovery of *Hutchinsoniella*.

With the coelacanth *Latimeria chalumnae*, we move into classic living fossil territory. Keith S. Thompson gives us a well-balanced history of the discovery of the living members of this ancient group of fishes. Long known from the fossil record, coelacanths had seemingly become extinct at the end of the Cretaceous. Thompson draws attention to the ecology of *Latimeria* (to the rudimentary degree to which it is known)—an interesting theme given the growing conviction that survival, every bit as much as adaptive change, is mediated by environmental stability and change.

Arthur W. Galston rounds out the case histories by recounting the fascinating story of *Kakabekia*, thereby in one blow expanding the cases to include plantlike microorganisms, and ranging back as far as the mid-Precambrian. At two billion years, *Kakabekia* is a truly prodigious case of anatomical, and presumably physiological, conservatism.

But why, in general, do we see these cases of extraordinary evolutionary slow motion? In the fourth essay, I try to grapple with underlying patterns of similarity that link up most, if not all, of the classic cases of living fossils. I conclude that many examples involve species that are ecological generalists, able to survive life's vicissitudes because they have not specialized too closely on any particular aspect of their energy resources or physical environment. The theme reappears immediately in part 6, on extinction, in which avoidance of specialization is implicated with resistance to extinction. But why this lack of anatomical change along the way? I argue here that ecological specialists belong to lineages that speciate rapidly, offering the opportunity to acquire still more anatomical novelties. Closely related specialist species tend to live in the same area to a much greater extent than more generalized species—suggesting that specialists divide up the ecological resource pie, while generalists tend to exclude one another from the same area. Thus interspecific competition plays a big role in the theory I present in part 5. Given the current state of ecological thinking on competition (recall the article by Wiens in part 3), I am more inclined now to ascribe lower speciation rates in lineages of ecological generalists to causes other than pure interspecific competition. Readers interested in pursuing the topic of living fossils might consult a compendium of case histories edited by me and Steven M. Stanley, entitled, appropriately enough, *Living Fossils* (New York: 1984, Springer Verlag).

17

New Light on the Crustaceans

HOWARD L. SANDERS

Most biologists would agree that, after almost two hundred years' systematic collecting of animals in every imaginable corner of the earth, very few *major* morphological types can have eluded the taxonomists, and that such unknown major forms as may exist are to be discovered, in all probability, among the inhabitants of the relatively inaccessible great depths of the oceans.

This probability has been very recently verified by the discovery of the new phylum, Pogonophora, and of *Neopilina* (*Natural History*, March 1958); but in 1954, another taxonomic find was made—yet this time it occurred not at the great depths of the open sea, but on the bottom of the hardly inaccessible waters of Long Island Sound. This new animal proved to be a rather peculiar crustacean.

The crustaceans—which include crabs and lobsters as their best-known representatives—are often viewed as the aquatic counterparts of the terrestrial insects. Both groups, together with the spiders, millipedes, centipedes—and the long-extinct trilobites and eurypterids of Paleozoic times—belong to the phylum Arthropoda. They are animals with segmented legs and a hard, chitinous exoskeleton. The single phylum Arthropoda contains *more* animal species than the combined total of the rest of the animal kingdom.

The class Crustacea, within this phylum, is divided into six major categories formally known as subclasses. And because each subclass constitutes a sharply demarked unit without intermediate forms, the relationships of these subclasses to one another and the history of their origin are poorly understood. For this reason, the theories that have been proposed to explain crustacean evolution are speculative and controversial.

We are all familiar with crabs and lobsters, and perhaps less so with sand fleas and sow bugs. These are all representatives of a single crustacean subclass, the Malacostraca. Copepods, those minute organisms that constitute a large part of the zooplankton found in both fresh and salt walter, form another crustacean subclass, called, not surprisingly, the Copepoda. The highly modified barnacles, found encrusted on rocks at low tide, represent still another group, the Cirripedia. A fourth category—the Ostracoda—includes the small, shelled crustaceans which are reminiscent of tiny clams. The brine shrimp, the fairy shrimp and *Daphnia,* the water flea—often possessing a large number of body segments—are members of the fifth subclass, the Branchiopoda. The sixth subclass, the Mystacocarida, discovered as recently as 1942, are a small group of crustaceans that live in the spaces between sand grains in intertidal beaches.

In the process of analyzing samples of bottom-dwelling animals from Long Island Sound in 1954, I came upon a number of small, elongated, somewhat transparent crustaceans. These minute creatures, hardly a tenth of an inch in length, were strikingly different from any of the crustaceans with which I was familiar, for a cursory examination revealed a large number of body segments whose presence excluded the animal from all the crustacean subclasses except the Branchiopoda. Yet, the bifurcated (or biramous) trunk limbs—each with an inner and an outer branch—were markedly different from the leaflike structures characteristic of that group. It seemed possible, then, that this organism represented a new taxonomic type.

Subsequently, more detailed studies of the tiny creature's morphology have verified this initial conjecture, leading me to erect a seventh crustacean subclass, the Cephalocarida, to include this animal—which I have named *Hutchinsoniella macracantha,* in honor of Professor G. E. Hutchinson, of Yale University. As taxonomic nomenclature may appear puzzling to the uninitiated, it should perhaps be said that these two terms were not chosen at random. The subclass name, Cephalocarida, stems from the fact that these animals have a covering or carapace over the head that is reminiscent of a shield; hence the term "head shield." The word macracantha, in turn, refers to the extremely long spines, pointing backward on the animal's telson, or tail.

Even more exciting than the creature's uniqueness, however, were indications that it shared certain morphological characteristics with the more primitive members of most of the other crustacean subclasses—for, as it was pointed out earlier, the absence of such suggestive, intermediate forms has in the past been the chief barrier to fruitful speculation regarding the evolutionary history of the crustaceans. Now, in *Hutchinsoniella,* the trunk limbs seemed most similar to those found in the Leptostraca, a primitive order of the subclass Malacostraca; yet the head appendages are almost identical to the same structures in the more generalized or larval representatives of the copepods, barnacles, and ostracods. Finally, the new animal's numerous body segments and the paired ventral nerve cord are uniquely shared with branchiopods. What all these resemblances seem to suggest, then, is that *Hutchinsoniella* might be a long-sought-after "missing link."

There is circumstantial evidence for such a surmise in the fact that the *adult Hutchinsoniella* has taxonomic features present only in the *early* (or naupliar) stages of the development of many other crustaceans. For example, *Hutchinsoniella* retains the so-called masticatory process—a small extension at the base of the second antenna, used for feeding—until a late juvenile stage. This same structure is found only in the naupliar stage of other crustaceans. Furthermore, the median or so-called "naupliar" eye in the adult *Hutchinsoniella* is on the ventral rather than dorsal surface, which agrees with its position as it is found in the nauplius of other crustaceans.

In fact, it is tempting to regard our *Hutchinsoniella* as merely an elongated adult nauplius. From such "ancestral" stock, it would be possible to derive the other crustacean subclasses by the process known as neoteny. In this process, an organism becomes sexually mature at an earlier developmental, or larval, stage— the later stages are sloughed off and are forever lost from the genetic constitution of the species. Unencumbered by the more specialized adult characteristics, the organism becomes genetically plastic and can respond readily to its environment.

Such a hypothesis envisions diverging evolutionary patterns, initiated perhaps several times as a response to neoteny, eventually resulting in the formation of a number of the present-day crustacean subclasses. The scheme seems to make sense, since only the early larval or naupliar stages are common to all the crustacean subclasses, while there are few morphological similarities among them in the later stages, except those which are shared with the new subclass, the Cephalocarida.

What can we learn from the fossil record to gauge the relative significance of this new crustacean find? Crustacean remains from the Paleozoic period— approximately 250 to 500 million years ago—are few. Of those fossils that are definitely demonstrated as crustaceans, none appears to be as "primitive" as this small, contemporary cephalocarid from Long Island Sound. Apparently, the closest fossil relative of *Hutchinsoniella* is the wonderfully preserved fossil from the Rhynie Chert deposits of Scotland, known as *Lepidocaris rhyniensis,* which lived about 300 million years ago. Yet this mid-Devonian relation—already an unmistakable branchiopod with the reduced head appendages characteristic of this subclass—is more specialized than is our morphologically conservative *Hutchinsoniella.* Rather, using the trunk appendages of the Rhynie Chert crustacean as an *intermediary,* it seems perfectly possible to derive both types of crustacean trunk limbs—the biramous and foliaceous—from the cephalocarid appendage, as we find it today on *Hutchinsoniella.*

Thus, the evidence from three sources—comparative anatomy, embryology, and paleontology—points strongly to the conclusion that *Hutchinsoniella* represents a life form with a history of almost no change over a period of hundreds of millions of years. Moreover, when this modern cephalocarid is compared to various hypothetical crustacean ancestors that have been postulated by different students, it is remarkably similar to their imagined reconstructions.

Accepting the ample evidence that *Hutchinsoniella* is in all probability the most primitive known crustacean, what can the animal provide in the way of clues to the relationship of the crustaceans to other members of the phylum Arthropoda?

The most primitive class of arthropods was, undoubtedly, the Paleozoic trilobites, extinct for at least 200 million years. In the trilobites, the appendages on both the head and trunk were, except for the first pair of limbs, identical. These appendages were biramous, that is, the leg gave rise to an inner branch (endopodite) and an outer one (exopodite). The endopodite had seven joints, and the segment making up the base of the leg had spinous lobes, called gnathobases, on its inner margin, which were used in feeding. Each of these limbs was used both for feeding and locomotion, and no appendage was specialized for any specific function. Certain of the same limb components can be demonstrated for the spiders, horseshoe crabs, and the eurypterids, and it is now quite widely accepted that all these groups were derived from this initial trilobite stock.

Now, there exists a provocative and remarkable assemblage of fossils from the mid-Cambrian Burgess Shale deposits of British Columbia, 450 million years old. The arthropod fossils of these deposits are strikingly diverse in appearance. Some are typical trilobites; others could be included in the group that gave rise to the modern Xiphosura, represented by the horseshoe crab; while still other individuals are markedly crustacean in appearance. All the varied forms have one important feature in common; they possess the characteristic trilobite limb and the trilobite appendage pattern. Primarily for this reason, even the Burgess Shale "pseudocrustaceans" have been placed within the class Trilobita.

Crustacean antecedents, however, are still poorly understood. Until fairly recently, it was generally believed that the trilobites were also the ancestors of the Crustacea. At present, the prevailing concept—largely as a result of the detailed studies of the Norwegian paleontologist Stormer—is that the crustaceans and trilobites are *not* related.

In fact, some even believe that both groups were evolved independently from different nonarthropod ancestors. They argue that neither the generalized limb series found in the trilobites nor the trilobite appendage can be detected in any known member of the Crustacea, even in the Paleozoic forms—although the gnathobases we see in trilobites do occur on the trunk limbs of the branchiopods. Instead, they point out, crustacean limbs are characteristically specialized for different specific purposes in different regions of the body. Those on the head have either a feeding or a sensory function, while the limbs of the trunk region are utilized exclusively for swimming or crawling.

The new-found *Hutchinsoniella,* however, possesses some remarkable noncrustacean, trilobite characteristics. All appendages, except the first, are used both for obtaining food and for locomotion and, contrary to all other Crustacea, no *Hutchinsoniella* limb is specialized for a specific function. This is precisely the pattern that was present in the trilobites.

Furthermore, the limbs of *Hutchinsoniella* are structurally similar: that is, no appendage is very markedly modified from a basic common plan. In fact, the last head limb (second maxilla) is identical to the trunk limb series. No such clear-cut homology of parts can be demonstrated among the appendages in any other crustacean, nor is there a case where a head limb bears a structural resemblance to the trunk appendages. Finally, *Hutchinsoniella,* alone among the Crustacea, retains what seems to have been universally present in all trilobites, a seven-jointed endopodite.

All this, I feel, shows clearly that the trilobites and the crustaceans *must* have been related, although it is difficult to determine whether the crustaceans evolved from the trilobites or whether both groups were derived from a common ancestor. It is just possible that the Burgess Shale "pseudocrustaceans" may represent a precrustacean stage of evolution which could have served as precursor to the crustacean stock.

In any case, it seems clear that the primitive arthropod limb pattern consisted of a series of identical appendages, with each limb serving at least a dual function. With the evolution of the Crustacea, there was a specialization of these appendages, with the loss, on the head limbs, of the component used for locomotion and the concurrent loss of the feeding parts of the trunk appendages. Our new addition to the ranks of the animal kingdom, *Hutchinsoniella,* seems to represent an arrested, early stage in this ancient line of evolution, since its limb pattern is intermediate between the original arthropod plan and the scheme found in all other crustaceans. Of course, the phylogeny of the Crustacea still leaves many and vast questions to be resolved; but it would seem to be one of history's characteristic ironies that so tiny a creature as *Hutchinsoniella* has helped us to pose them properly.

18

Secrets of the Coelacanth

KEITH S. THOMPSON

LIVE COELACANTH CAUGHT THIS AM. MOVIES DISSECTION
OK. STOPOVER PARIS. HOME AROUND THE FIRST.

This cable, sent in 1972 from the Comoro Islands by Robert Griffith of Yale, marked the latest success in research on the coelacanth, the fish that scientists once thought had been extinct for millions of years. For the first time a coelacanth had been caught, kept alive for several hours, and studied in detail. Precious tiny cargoes of frozen and chemically preserved tissues have since been sent to scientists all over the world, and research on the coelacanth is now proceeding at an intense level. This fish, always so enigmatic, so difficult to catch and study, is slowly yielding up its secrets. But there is still much to be discovered and the direction of future research is not clear.

The first coelacanth known to science was a small fossil wrongly described as a bowfin in 1822. Thereafter, several types of coelacanth were correctly recognized, and in 1855 Louis Agassiz gave the group its formal name Coelacanthini, after the Carboniferous genus *Coelacanthus* ("hollow spine"). At present, there are twenty-eight known fossil coelacanth genera, ranging in age from the Early Devonian to the Late Cretaceous. Since no fossil has been found of more recent age, it was thought that the group became extinct 70 million years ago, along with most other fauna of the Age of Reptiles.

Coelacanths have always interested paleontologists because of their morphological similarity and phylogenetically close relationship to the lungfish and to the Rhipidistia, a Paleozoic fossil order. Together these form the "lobe-finned fishes," and by the turn of the century it was clear that it must have been from within this group (actually from the extinct Rhipidistia) that the first amphibians evolved in Devonian times. For all their similarity to the ancestors of the tetrapods, however, it was clear that the coelacanths were a largely marine group of fishes. To the student of tetrapod origins, the air-breathing lungfish, with three living tropical fresh water genera available for study, were always of more immediate concern in the search for analogies with the Devonian Rhipidistia and amphibian origins.

But then, in December, 1938, a living coelacanth was trawled from the bay of the Chalumna River, near East London, South Africa. It was an event of spectacular significance, worthy of the *Lost World* of Sir Arthur Conan Doyle. The late professor J. L. B. Smith of Grahamstown University identified the fish, and published a scientific description of it, although all that remained when he saw it was the skin and part of the skull. Smith named it *Latimeria chalumnae*, after Miss Courtenay Latimer of the East London Museum—to whom the fish was first taken and who called in Smith—and after the place of capture.

For fourteen years the search for further specimens continued, until in December 1952, a second specimen was brought to Smith's notice. This one had been caught on the island of Anjouan in the small Comoroan group to the northwest of Madagascar. A local fisherman had recognized it from a poster, which Smith had distributed widely up and down the east coast of Africa. He claimed the reward, and Smith, in a plane loaned to him by the South African prime minister, flew to the Comoros in order to collect the specimen.

The Comoro Islands are French territory, and since 1952 intensive research into the coelacanth, principally in terms of descriptive morphology, has been conducted at the Museum National d'Histoire Naturelle in Paris. By 1960 some forty to fifty specimens had been caught, preserved in formalin, and shipped to Paris for study. A long series of notices on the structures of *Latimeria*, together with the first two volumes of a planned comprehensive anatomy, has been published in Paris and elsewhere since 1953.

Regrettably, of all the specimens captured before 1972, only one lived long enough to be observed by scientists. In 1955 a specimen was held in a submerged dinghy for some hours before it succumbed. About 1962, further specimens caught in the Comoros were offered for sale to museums around the world, and formalin-preserved specimens have now been distributed to Britain, the Soviet Union, North America, and Australia. From this worldwide effort a clear picture of the fish's anatomy has emerged.

Latimeria chalumnae is a large fish with a mouthful of sharp teeth. It reaches a length of six feet and a weight of 175 pounds. Its scales and fin rays are armored with many short, sharp spines. Its color is a muddy gray-blue, which rapidly becomes more brown after death. The eye is clear, but in specimens brought up from the depths it becomes cloudy because of pressure effects on the crystalline structure of the lens. The eye reflects a yellow color from a membrane over the retina. The skull has a curious intracranial joint, and the paired fins are large, lobed, and extremely mobile, with their own internal skeleton and muscle system.

There are no true lungs, but the fish has a fat-filled swim bladder evidently derived from the paired lungs of a distant ancestor. Thus, the fish cannot breathe air. There is no internal nostril. The stomach has a thick muscular wall, which often results in the contents being voided during capture, thus depriving us of information about the fish's diet. The rest of the internal organs need no mention here except for the reproductive system. Two gravid females have been caught, including one in 1972 that contained nineteen eggs of enormous size, more than 4 inches across. They lack any protective shell or membrane.

Apart from the first specimen, which came from off the coast of South Africa, *Latimeria chalumnae* has only been caught off the Comoro Islands and in fact from only two of these—Anjouan and Grande Comore. They have been caught at depths from 100 to about 800 meters (but mostly above 400 meters). More than seventy specimens have been taken at a capture rate of two to four per year.

To this bare catalog we could add a great mass of detailed anatomy, but there is a notable lack of information about the ecology and general way of life of these fish. This last affects not only our attempts to interpret what we have so far learned about the fish but also questions regarding how rare the fish is and how it has survived.

One reason for our long ignorance about the mere existence of living coelacanths and our continued inability to learn their secrets is the remoteness of the Comoro Islands. Set in the Mozambique Channel, the *archipel des parfums* consists of four volcanic islands with small fringing reefs. Beyond the reef, the underwater profile is very steep, plunging down some 2000 to 3000 meters. The geologic history and submarine topography of these islands may provide part of the explanation of how the coelacanth survived from prehistoric times. Two hundred million years ago coelacanths were a not uncommon part of shallow sea faunas worldwide. Is it a coincidence that some of the best-known fossil coelacanth specimens come from the Triassic beds of nearby Madagascar?

There seems to be little to distinguish the Comoros from other islands in the western Indian Ocean, such as the Aldabra series. An expedition sent by the Royal Society of 1969 to some of these other islands showed that the fish populations on their deep underwater slopes are almost exactly like those of the Comoro Islands (although no coelacanths have been caught there).

Examination of the available statistics on catches shows that, apart from the first South African specimen, all coelacanths have been caught off Grande Comore and Anjouan, with a majority from Grande Comore. They are principally caught on the east coasts of these two islands and not at all from the other Comoroan islands of Mayotte and Moheli. The captures all come in the early hours of morning darkness. The majority are caught from January to March, but a large number are taken during the rest of the year. To interpret these data, we must know much more about the nature of the fishing effort. One may be reasonably sure, for instance, that the reason for their apparent absence from Mayotte and Moheli is the lack of extensive fisheries there. Could this also explain their apparent absence from all the other islands of the western Indian Ocean?

The Comoroan fishermen work alone in outrigger canoes hollowed from mango logs. They fish with a single hook on a vertical line sent down to the bottom by a chunk of rock secured by a slipknot, which is then released. They fish at, or very near, the bottom, normally in depths of up to 400 meters. Presumably this is the deepest that can conveniently be worked in this fashion. Curiously, the months from January to March present the very worst weather in the Comoros.

One needs to ask, therefore, why they bother to fish at this time, at night, and at these depths. The answer is that they are not normally seeking the coelacanth itself, but rather food fishes, especially the oilfish *Ruvettus pretiosus,* which is relatively abundant at these depths around the islands, reefs, and banks in the western Indian Ocean. In the process they catch coelacanths, apparently largely by accident. Now when they are asked to go after a coelacanth, they obviously go to where they have previously caught them—the oilfish grounds. Thus, to put an extreme interpretation on the catch data, we might say that they present the coincidence of oilfish and *Latimeria* distribution at the season when oilfish are sought. This may actually be a bit strong, but at least it seems clear that the catch data reflect only the fishing effort and that fishing is not directed principally toward the coelacanth itself.

Given this unsatisfactory information, what can we say about the distribution and numbers of the fish? The sex ratio in the catch is approximately 25 males to every 20 females. The size range has remained more or less constant, and no fish of less than 34 inches (estimated to be at least four years old) has been taken. This suggests that the catch rate (steady for nearly twenty years) has not significantly depleted the stocks. It seems likely that there are coelacanths around the other Comoroan islands that have not been taken at all. At this point, to develop an interpretation, it is necessary to turn back to the evidence from the dead fish.

Latimeria does not seem to be an open-water fish. The heavily armored surfaces and specialized lobe fins, as well as the nature of the other fins, suggest constant association with a hard substrate. While the caudal fin is a powerful structure, it is obviously only adapted for short, rapid bursts of swimming. The paired fins and the second dorsal and anal fins are adapted to slow, sculling movements by which the fish presumably moves along or near the bottom, pushing off from hard objects with the paired fins, waiting for prey to appear, and then taking it in a quick lunge.

The possibility is strong that coelacanths engage in a nocturnal migration toward the surface and that by day they lie at greater depths. The eye is adapted only to low light intensity, the properties of the visual pigment comparing best with those of fishes known to live in deeper waters. The few stomach contents recorded include lantern fishes, which are known to be vertical migrators.

The fats in the swim bladder, and dispersed widely in the oily tissues of the coelacanth, contain many very light wax esters and presumably are an adaptation for giving neutral buoyancy. There are indications, its color, for example, that *Latimeria* is not a truly deep-sea fish. We might reconstruct, therefore, that *Latimeria* is a fish that lives a relatively sedentary existence near the bottom. At night it may come toward the surface to follow prey species, and migrate back down the submarine slopes of the islands during the day. If this is correct, then apart from detailed ecological preferences that we do not know of, the species could be widely dispersed within the area immediately around the Comoros. But to move beyond this region, the fish would probably have to swim in open water, away from the bottom, which may explain its restricted distribution. As to population size, we cannot guess, but common sense tells us that to be safe we should avoid a catch rate higher than the present one.

There is a definite limit to what can be learned from museum specimens chemically preserved in formalin. Therefore in 1966 an attempt was made by the Comoroan authorities to freeze and ship a fresh specimen to Yale. The venture succeeded beyond all our hopes, and on Memorial Day, 1966, a group of scientists gathered around a laboratory table and made the first detailed examination of a fresh coelacanth. Within a few weeks it was clear that surprising results would be obtained. Prof. Grace E. Pickford of Yale discovered that *Latimeria,* unlike all other known bony fishes, balances the osmotic pressure of its blood by a mechanism of urea retention generally identical to that used by the elasmobranch fishes (sharks and rays). Prof. G. W. Brown and Dr. Susan Brown discovered that the liver contains all the enzymes of the ornithine cycle—the biochemical process by which the urea is produced. I found that, contrary to widely held opinion, the skull of the coelacanth, which is characterized by a set of joints separating the braincase into two parts, is movable through some eight to ten degrees, producing one of the most curious mechanisms of jaw movement known in vertebrates. Since this mechanism is also found in the fossil Rhipidistia, it was most important to discover what happens in a live specimen.

From the Yale coelacanth, "fresh" specimens of a wide variety of tissues were rushed to biologists at every hand, all of whom wanted a chance to compare the biology of this "primitive" fish with what is known of other living fishes. Interest in the biology of ancient fishes and the origin of tetrapods rose to a new high. But with the study came problems. Obviously more material was needed, the history of which was known exactly. The study of frozen tissues pointed up a whole series of exciting problems (what, for example, was the visual pigment of the eye like?) that could only be answered in one way—we had to get our hands on a completely fresh specimen.

In January 1972, a joint expedition of the Royal Society (London), the Muséum National d'Histoire Naturelle in Paris, and the National Academy of Sciences (sponsored by the National Geographic Society) arrived in the Comoros. The expedition had been planned for nearly five years. Biologists all over the world had been canvassed and a complete set of protocols for optimum preservation of tissues from a coelacanth had been set up. Every effort would be made, should a specimen be caught, to keep it alive as long as possible in order to make observations on the living fish. The party planned to use every possible fishing method. They had their own gear, to be used from a launch, and also intended to mobilize the local fishermen.

In the end it was, as ever, only the local fishermen who succeeded. In fact, the entire success of the expedition resulted from the labors of the Comoroan fishermen, who went out night after night to help the visitors in their search for the *gombessa djomole.*

The expedition was immediately successful. On the night of January 6, before the American contingent had even arrived, the British and French got the news that a large specimen had been caught off Anjouan. But by the time the scientists made the 70-mile journey from Moroni, the fish was dead. Much useful information, however, including priceless tissue samples, was obtained. The specimen was only the second mature female to have been caught and it contained nineteen enormous eggs, perhaps the largest fish eggs ever seen.

Bad weather intervened, but the fishermen were persuaded to redouble their nightly work. By mid-March a familiar pattern had set in: rainy days, sunny days, lots of fishing, no coelacanths. Only Robert Griffith of Yale and Adam Locket of the Institute of Ophthalmology in London remained when at 3:00 A.M. on March

22, Mohammed Ali Chabane, the headman of the village of Iconi on Grande Comore, woke them up—a live coelacanth had been caught at Iconi, 12 miles away.

The fish had been caught by Madi Youssouf Kaar and brought to shore, where it was transferred to a submerged wire mesh cage. Griffith and Locket observed it by artificial light until daybreak, making detailed notes on its color and behavior, especially the use of the fins. They then transferred the fish to a large tank where they made a motion picture record of it. The reconstruction of the swimming of coelacanths is based largely on their observations.

One cannot be sure why coelacanths die when brought to the surface: it must be a combination of decompression effects, temperature change, and general shock. Lacking specialized equipment to keep it alive (for this was only a pilot expedition), the two scientists prepared to dissect their prize. The specimen, an immature female only 33 inches long, had been caught only 600 meters from shore at a depth of about 100 meters. It lived for a total of seven hours after capture, most of the time confined under difficult conditions.

The 1972 expedition finally established that closely supervised intensive fishing by the local fishermen can confidently be expected to produce a coelacanth if one is prepared to wait a couple of months. It confirmed that specimens can be kept alive for a while and raises the hope that more elaborate methods might succeed in keeping one alive longer. A great deal of research material was gained, and the osmotic physiology, blood, eye, reproductive system, bone and cartilage, and many aspects of the basic biochemistry of the coelacanth are now under scrutiny. Observation of the live fish showed something of the swimming and respiratory system. Now, what should be the pattern for future research?

Among the results gained by the expedition was the capture of a mature female. The nature of the eggs indicates that some kind of special breeding behavior occurs in the coelacanth. It may be ovoviviparous (a view Dr. Griffith and I favor) or, at the very least, if the eggs are laid in the water, close parental care must be shown. Therefore, whatever the size of the coelacanth population, there may be considerable danger in random fishing efforts during the breeding period.

In 1966 I wrote, "Perhaps the next phase of study of the coelacanth should be conducted with the investigator holding a motion picture camera rather than a scalpel." I now consider this to be even more pertinent. When the results of laboratory studies now in progress are known, there will undoubtedly be a need for a second expedition to obtain a similar specimen.

But our need to know the ecology of the fish will not be met by this process. I am of the opinion that no random, or even systematic, fishing program will suffice. The present catch rate is too low, and even if one could devise a fishing technique that would allow a large catch to be made in a systematic survey of the area (which seems unlikely), the chances are too strong that the fish would be made extinct in the process. The only safe investigation of the general biology of this fish can come from direct observation. A systematic survey of the Comoroan deep slopes by means of photography, television, and observers in a submersible would, at the very least, tell us something about the physical and biological environment, even if no coelacanth were seen. A film record of a live coelacanth in its natural environment would be worth more than fifty specimens on laboratory tables, and I would certainly rather see the effort go in this direction.

There are, of course, also considerable legitimate pressures for the capture of a live specimen and its preservation in an aquarium. It would be a marvelous step for the science of biology, as well as for the public at large, if one could be put on general view. At the moment we do not know under what conditions the fish normally lives and can only guess at why it dies when brought to the surface, but the extreme stress of capture must be very important. I believe that the correct procedure would be to prepare a suitable tank, in which at least the parameters of pressure and temperature could be controlled, to install this tank on Grand Comore, and to instruct local officials in its operation.

In fact, such an operation could be set in progress without the enormous expense of a major scientific expedition. For example, mobilization of the local fishermen seems likely to produce one or two specimens per season, and better communication between the fishermen and the shore could improve the chances of a specimen being kept alive after capture. A fast launch summoned by radio could quickly bring the fish back confined in a simple tank of seawater. This would be much better than the present long paddle home with the fish towed behind, a hook through its head.

Study of the coelacanth has had a long and often frustrating history. Some thirty-five years after the discovery of the living *Latimeria chalumnae,* we are beginning to assemble a large set of data concerning the biology of this enigmatic fish. But a fundamental aspect of its biology—its ecology and general way of life—remains almost a complete mystery, enlightened only by inferences made from the specimens all caught at random by Comoroan fishermen. Such specimens have been studied with great success, but new methods, including the maintenance of specimens in aquariums and direct underwater observation of the fish and its habitat, will be needed before we can truly understand the biology of this curious relict of Paleozoic and Mesozoic times.

19

A Living Fossil

ARTHUR W. GALSTON

Imagine the emotions of a paleontologist, long a student of dinosaur evolution, who suddenly encountered a live *Brontosaurus, Triceratops,* or *Tryannosaurus.* That is roughly what happened about ten years ago to Sanford Siegel, a University of Hawaii botanist, when he examined the microorganisms of a sample of soil he had gathered near the wall of Harlech Castle in Wales. When he cultured that soil sample in the presence of concentrated ammonium hydroxide, which greatly inhibits or arrests the life processes of most conventional cells, the medium triggered the growth of microscopic clusters of star-shaped bodies attached to slender stalks. Each body, about 5 micrometers (0.0002 inch) in diameter, closely resembled pictures Siegel had seen of a recently discovered fossil microorganism. But as far as he knew, no living specimens of this organism had ever been described. With the help of the fossil's discoverer, Elso Barghoorn of Harvard University, Siegel was able to establish, in a strange sequence of paleobotanical events, that he had found a living relative of an organism first described as a fossil.

Barghoorn had made his own discovery while gathering specimens of ancient rocks in a search for primitive organisms. One specimen of chert, or flintlike rock, from Kakabek in Ontario, Canada, contained peculiar umbrellalike forms that seemed regular enough in physical appearance and structure to be considered microorganisms, rather than a pattern that had developed as the rock formed. Barghoorn named these microorganisms *Kakabekia umbrellata,* meaning umbrellalike form from Kakabek.

Since the rocks in which Barghoorn's forms appeared dated from the middle Precambrian period, about two billion years ago, the microorganisms were among the oldest of all plantlike fossils. Siegel's discovery of a living relative of Barghoorn's fossils established a remarkable thread of biological history. Siegel named his creature *Kakabekia barghoorniana* in honor of his colleague.

Siegel had come to Harlech Castle in the course of a long project. For many years, with the support of the National Aeronautics and Space Administration, he had been examining the physiology of organisms under stress, especially stress caused by harsh environments likely to be encountered during space travel near or on other planets. Because ammonia is generally thought to have been one of the more abundant components of the primitive earth's atmosphere, and is still one of the major components of the present atmosphere of Jupiter, Siegel wanted to study earthly environments with abundant ammonia. Natural candidates for examination were soils saturated with urine. On a chance visit to Harlech Castle, he observed tourists urinating near the castle walls, and learning that this was an old practice, decided to collect his first soil sample. Since the organism he discovered in that soil thrived in the presence of concentrated ammonium hydroxide and was not seen in other environments, Siegel hypothesized that he had found an obligate ammonophile, an organism that requires ammonia in order to grow.

In the ten years since he first visited Harlech Castle, Siegel has found that he was mistaken about *Kakabekia's* need for ammonia. At sea level, the microorganism does require the compound, yet *Kakabekia* also appears in certain mountainous regions low in ammonia but high in alkalinity. Without ammonia, *Kakabekia* needs special soil conditions, as well as peculiar combinations of temperatures and altitude, to thrive.

In soil samples from Hawaii, California, the Great Plains states, Illinois, New York, western Europe, and northern South America, Siegel found no signs of *Kakabekia*. But the microorganism did show up in soils from Alaska, Iceland, and various alpine regions. So temperature seemed to be one of the factors governing the distribution of modern *Kakabekia*. But in studying its distribution up Hawaiian and Japanese mountain peaks, Siegel found that it grew in bands, which meant that at least one other factor was interacting with temperature to limit the organism's distribution. At lower elevations, *Kakabekia* is not found below about 45° north latitude, and thus appears restricted to regions with comparatively cool summers. At latitudes closer to the equator, it grows only at altitudes above 6500 feet. Since air temperature decreases about 7°C (about 12°F) for each 3300 feet of altitude, *Kakabekia* seemed to require low temperature and a certain altitude.

The *Kakabekia* Siegel found on Mauna Kea, an extinct volcano in Hawaii, appeared to bear out his ideas about the microorganism's temperature and altitude requirements. He recovered it on the volcano at altitudes above 11,500 feet. Once cultured, organisms consistently appeared in his soil samples at − 7°C, but not at

30°C. But at about 10,000 feet up Mauna Kea, and on down to the base of the mountain (about 7,500 feet above sea level), Kakabekia showed exactly opposite characteristics, growing at 30°C, but not at − 7°C. This second Kakabekia, which has adjusted ecologically to warm temperatures, seems to be a variant of the cryophilic, or low-temperature-favoring, strain that Siegel had found earlier.

There are other variants of Kakabekia, with differences in the umbrellalike cap, with or without a stipe, or stalk. The umbrella may be lobed, scalloped, fringed, or cut into rather acute angles, and may have varying numbers of rays, the umbrella's ribs. These strains probably represent genetic variants that evolved from some prototypical form.

Another peculiarity of Kakabekia is that its cultures require no oxygen, but unlike typical anaerobic bacteria such as Clostridium, which causes gangrene, it is not killed by oxygen in the air. Kakabekia does require a distinctly alkaline environment, which can be furnished equally well by sodium hydroxide, potassium hydroxide, or ammonium hydroxide at a level that would be toxic to most living creatures. When sodium hydroxide is replaced by sodium metasilicate, which provides the same kind of alkaline environment, as well as silica, Kakabekia grows more slowly, or stops altogether. Since it contains large quantities of silica (the hard part of its cell walls that leaves a fossil) and has almost always been found with diatoms (algae that have silica cell walls), researchers thought that Kakabekia could easily take advantage of silicon in its microenvrionment. But if metasilicate is toxic to Kakabekia, it must obtain its silicon from ordinary minerals in soil, possibly from sand, which is silicon dioxide.

So far we know little about the Kakabekia cell and its metabolism. Our information has been limited because we have been unable to grow Kakabekia in pure culture—all cultures have been contaminated by other organisms, invariably diatoms—and because Kakabekia cells seem to grow very slowly. We do know that when these cells are tested with the Feulgen stain, which shows up the DNA of chromosomes, the results are negative. Since all known living things must contain DNA, Kakabekia is probably a prokaryote, an organism that, like a bacterium, does not have its chromosomes gathered into a discrete nucleus. This hypothesis is supported by the discovery by several Russian workers of some star-shaped bacteria that develop in certain kinds of soil or in creek water rich in organic material. Although Kakabekia is nearly twice as large as these bacteria, they do have a marked morphological resemblance. Their possible genetic relation is seconded by the fact that all of the star-shaped bacteria have been collected from cool regions.

Kakabekia's slow growth raises questions in itself. In order to grow, it must have oxidizing enzymes that will mobilize energy from its environment. Yet preliminary experiments have revealed that *Kakabekia* contains neither the deme enzyme nor the phenol-oxidizing enzymes that are most organisms' conventional means of producing energy.

At present, researchers' questions about *Kakabekia* far outnumber the answers. Siegel and other workers will be looking for the microorganism for some time in such scattered places as Point Barrow, Alaska; Surtsey Island off Iceland; the top of Mauna Kea in Hawaii; and Harlech Castle in Wales. When all of the living fossil's long history is clear, microbiologists and evolutionists will have new stories of their own to tell us.

20

Survivors from the Good Old, Old, Old Days

NILES ELDREDGE

The most unchanging aspect of life is change itself. Yet a number of animals, known as "living fossils," have persisted essentially unchanged for hundreds of millions of years. Such an animal is the North American horseshoe crab, *Limulus polyphemus,* which occurs commonly along the eastern seaboard of the United States and around the shores of the Gulf of Mexico. This creature is not a true crab; it is the sole modern representative of a group, distantly related to spiders and scorpions, that was abundant in the Paleozoic seas. Some other living fossils are the nut clam, *Nucula;* the brachiopod *Lingula;* and the lizardlike reptile *Sphenodon,* perhaps the most famous vertebrate living fossil.

To qualify as a living fossil, an animal must have certain credentials. It must represent the persistence, in nearly unchanged form, of an ancient and comparatively primitive structural type. It must have fossil relatives that date far back to a remote geologic period—as close as possible to the 600-million-year-old beginnings of the fossil record of complex animals. (Fossil horses, which are essentially modern in structure, exist in Pleistocene rocks that are more than a million years old, but a few million years just are not enough; the true living fossil's pedigree must be tens, or preferably hundreds, of millions of years old.) And a living fossil should have no—or at best, only a few—close living relatives. In addition, if the creature exists in a restricted area, so much the better.

Because animals with these qualifications raise many provocative (and largely unanswered) questions in evolutionary biology, they have long excited the interest of naturalists. Why have some organisms evolved so slowly, if at all, while the rest of life on earth has been involved in the cycle of evolution and extinction? Why do comparatively primitive structural types frequently seem to outlast their more advanced relatives, with the result that most classic living fossils have few, if any, even moderately close living relatives?

A structural type is simply a plan of body organization. All members of a species share a basic body plan. But "structural type" can also be applied to large groups of related organisms. All mollusks, for instance, share a basic, discernible body plan that differs drastically from the basic plan of organization of arthropods, vertebrates, and indeed all other known animal phyla. The old notion of the scale of nature—the gradation of organisms from simple to complex—was based on the comparison of body plans of the major animal groups. From this point of view, all the phyla are living fossils, but the simplest organisms—the unicellular protists, the tissueless multicellular sponges, and the simple-tissued coelenterates, for instance—are the best examples of living fossils because they are generally thought to represent a more "primitive" type of body organization than "higher" groups such as the mollusks, arthropods, and vertebrates.

One of the ironies of the early post-Darwinian era of evolutionary thought stemmed from translating the progression from simple to complex in the scale of nature into a direct progression of primitive to advanced, hence, of inferior to superior. The irony, of course, is the persistence of the presumably inferior body plans. If evolution only produces organisms superior to their antecedents, hasn't enough time elapsed for all the inferior primitive types to have disappeared, losers in the evolutionary race? But calling an entire phylum—like the sponges— a living fossil is also inexact, for we can readily imagine a large group persisting through the normal process of the extinction of its older members only to be replaced inevitably, as time elapses, by the evolution of newer species. Therefore, we should continue to restrict the notion of living fossils to species that closely resemble ancient species because it is on this low species level in the hierarchy of organisms that we must confront the phenomenon of evolutionary nonchange.

The importance of the qualification that living fossils be bereft of close living relatives is best illustrated by the disparate treatment usually accorded two modern clams, *Nucula* and *Neotrigonia*. *Neotrigonia* (meaning recent *Trigonia*) is one of the better-known examples of a living fossil. First known from an abundance of Mesozoic fossils, the trigonids—a diverse array of clams that made up an entire family—were thought to have disappeared from the fossil record at the close of the Cretaceous period. Long after they were discovered as fossils, living specimens of *Neotrigonia* were found along the southern coast of Australia. Somehow, one small group of these clams, whose heyday was over 90 million years ago, managed to hang on.

In contrast, *Nucula* is found burrowing into the mud of nearshore marine habitats virtually everywhere. Nuculoid clams, first encountered in Ordovician rocks that date back nearly 500 million years, are traditionally acknowledged to represent a more primitive state of "clamness" than the trigonids. Although trigonids have recognizable relatives that are perhaps 400 million years old, they did not become fully trigonid until the Mesozoic era and are a much younger group than the nuculoids. Nuculoid anatomy, as judged from external form and internal muscle scars and hinge teeth, has hardly changed at all from Ordovician times. Indeed, the shells of modern *Nucula* resemble those of Ordovician nuculoids more closely than today's *Neotrigonia* resemble Cretaceous trigonids. Yet *Nucula* is not cited as often or considered as good an example of a living fossil as *Neotrigonia*. Why not? Presumably because the abundance of individuals, the moderate diversity in terms of the number of extant species, and the broad geographic distribution of modern nuculoids disqualify this clam as a living fossil. *Nucula* is too successful, and a successful clam that happens to belong to a primitive group doesn't make a totally satisfactory living fossil.

Evidently, then, to be a good example of a living fossil, a creature should be a lone survivor, precariously clinging to its existence long after all its close relatives, unable to make it in more modern times, have died off. To this end, it helps if the organism has a relict distribution—that is, if it is found in only a small portion of the world, in a habitat whose extent is greatly reduced from the area of its former, far-flung stomping grounds.

Sphenodon, a reptile that lives exclusively on a few islands off the coast of New Zealand, meets all these requirements. This animal is the sole survivor of a moderately diverse Mesozoic order of relatively primitive reptiles. And, perhaps best of all, *Sphenodon* is confined to a single area that is vastly smaller than the worldwide range of its ancient relatives. It presumably hangs on to life only because there are no competing lizards in New Zealand.

The invertebrate organism that best fits this restricted set of criteria for a living fossil is the mollusk *Neopilina*. Discovered in the 1950s in dredge hauls from the abyssal depths of the ocean, *Neopilina* is the only living member of the Monoplacophora, a class of mollusks generally considered to be the most primitive. This class, which first appears in Lower Cambrian rocks, was thought to have become extinct by the end of the Middle Devonian period. The simple spoonshaped shell of *Neopilina* is virtually identical to that of the Silurian genus *Pilina,* a limpetlike reef dweller. but *Neopilina* does not cling to a reef in the intertidal zone. It exists exclusively on the deep floor of the ocean—a habitat exploited by a few other mollusks, indeed, few other animals of any class. *Neopilina* thus occupies a safe place free of competition from more highly evolved mollusks.

It is impossible to judge whether or not these primitive mollusks have always lived in deep ocean basins or whether they took a dive in the Silurian period, because no fossil record exists of deep ocean basin habitats older than the Jurassic. In any case, *Neopilina* admirably fits the most stringent requirements for a living fossil: a small group of closely related, primitive species of truly ancient pedigree that hangs on to life in the refuge of the abyss.

How can such basic types as *Neopilina* and other living fossils persist unchanged? Why do they tend to hang on for so long? Why do so many of these persistent organisms seem primitive when compared with their known fossil relatives? And, finally, why are these creatures frequently orphans without any close modern relatives?

In thinking about modern organisms, it is a truism to assume that all species are well adapted to their environments. In these days of ecoconsciousness, everyone knows that each species has its own niche, which consists of all the physical, chemical, and biotic influences—such as temperature, pH index, predators, and food resources—with which that species copes, utilizing its own peculiar behavioral and anatomical properties. When niches are thought of in this fashion, it is clear that no two individuals, let alone two separate species, share exactly the same niche characteristics. That much we know, but there is a tendency to forget that our lifetimes represent as thin a slice of geologic time as, say, seventy-five years in the Silurian period. So it is reasonable to assume that Silurian species were as well adapted to their niches as are we and our fellow creatures of the modern biota.

Niches are conservative, and species, if they persist at all, tend to do so unchanged as long as their habitats remain accessible to them. Given a habitat that persists almost unaltered over a long period of time and a species adapted to it, there are only three possible courses for the species to follow: it will either disappear, evolve into something new, or persist in a form similar to its original one. Instances of this latter kind of morphological stability, in which a species persists essentially unchanged for perhaps as many as five or ten million years are currently being carefully documented by a number of paleontologists. Such long-term stability makes one wonder why evolution was not basically over with soon after the first appearance of life on earth. In a sense, it was, since all of the phyla of the complex, multicellular animals that appeared near the beginning of the Cambrian period originated almost immediately. But evolution proceeds on the species level, and no living metazoan species (those having differentiated cells) inhabited the Cambrian world. Extinction has claimed the earliest metazoans; indeed, it has claimed the vast majority of all the species that have ever lived on this planet.

So far I have argued that if a species survives, it tends to do so unchanged. Also, the chances of a species persisting over the truly long periods of time required by living fossils are almost nil. But it happens that the chances for survival are not equally distributed among all species or even among different species that are closely related to one another and belong to the same order or even the same family. Some species are specialists, adapted to a highly specific set of conditions that they require for continuation, while other species are tolerant of a much broader range of environmental conditions. Evolution is essentially opportunistic, and wherever a set of optimal, stable, predictable conditions develops—for example, a seaway with little daily or even seasonal fluctuation in salinity, temperature, or wave action—the locale will soon be populated by a diverse array of species exploiting the subhabitats that such a favorable environment offers. These species frequently develop anatomical and behavioral modifications of their ancestral condition, the better to exploit the advantageous conditions of the habitat. Specialist species thus often look and act more highly evolved than their forebears and are commonly cited as evidence that evolution produces biological improvement.

Nonspecialist, or generalist, species, on the other hand, which occupy unpredictable environments with a wide variety of daily and seasonal changes, must be more flexible. They do not concentrate too much on one particular food item or get too used to a certain comfortable temperature. This kind of species is comparable to the person who is a jack-of-all-trades but master of none. *Limulus,* the horseshoe crab, is such a generalist species. It is a moderately good swimmer, walker, and burrower. While some true crabs are better walkers or better burrowers or better swimmers, I know of no crab that can perform all three functions as well as *Limulus.*

Generalists cannot stand too much competition, so they live in habitats where most other species are also compelled to be generalists. In addition, they do not show much diversity; that is, at most only a few closely related generalist species will be found living together at any one time and place. The horseshoe crab is a single, far-flung species, which ranges from Nova Scotia to the Yucatan. Its only close relatives—three or four species that are dead ringers for the North American species—live in different parts of the western Pacific and Indian oceans.

Competition between closely related, relatively specialized species frequently leads to a subdivision of resource space. Various species somehow manage to live side by side, dividing up the resources and concentrating on different aspects of the environment, such as different food items. In the process, their morphologies and behavior patterns become even more specialized. Darwin first noted this specialization phenomenon while observing the native finches on the Galapagos Islands. Generalists, on the other hand, tend not to develop new anatomical and behavioral characteristics in the process of speciation. Instead, they continue to look and behave like their ancestral species. Competition between closely related generalist species thus usually results in the replacement of one species by another. That is why generalists appear to evolve so slowly and to have a lower rate of speciation. The ecological strategy of generalists acts as a damper on speciation because relatively few generalists will occupy the same amount of habitat space that a larger number of specialist species can populate.

Given these arguments, what happens when the environment is radically altered? To use a human analogy, think of a highly educated specialist in, say, banking, and a jack-of-all-trades handyman. The first may earn $50,000 a year and be considered a success, while the other earns far less and may be deemed a failure. The banker lives in a generally predictable environment and can count on a regular paycheck; the handyman lives catch as catch can. This may go on for years, until, bang—a depression hits. The banker may then lose his job, but the handyman, while also hit, can hang on and continue to do pretty much what he has been doing all along. The analogy is not perfect, but there is a wealth of evidence from both the fossil record and from the observation of modern living fossils, to show that generalists—the "failures" of the moment—are, in fact, the long-term winners. Survival is, after all, what life is all about.

At this point we can say that relatively unstable and unpredictable habitats, which persist through time, should be populated by generalists. In the marine realm, by far the most unstable, varying, and even unpredictable environments are the intertidal and barely subtidal zones near the shore. They are subject to great variations in temperature, salinity, amount of dissolved oxygen, presence or absence of water, and a host of other factors. Conditions are far more stable below the wave base out on the continental shelf, where even the sediments on the ocean floor are seldom disturbed. It is therefore not surprising to find such ecological generalists as *Limulus* and other marine living fossils inhabiting the nearshore environment.

Not all living fossils are obvious generalists, however, and the ecological generalist argument is not universally applicable. The reptile *Sphenodon* and the mollusk *Neopilina,* for instance, live in refuges safe from competition and possible extinction by less primitive and perhaps more efficient organisms, especially close relatives such as lizards and gastropods, respectively. The case of *Sphenodon* is apparently simply an instance of a favorable habitat; the case of *Neopilina* is a little more complex, since this organism had to become adapted to the harsh, rigorous environment of the deep ocean floor. But most examples of living fossils do seem to be ecologically generalized species when compared with their closest known living relatives.

Why should comparatively primitive plans of body organization lend themselves to a generalist survival strategy? Actually, they don't necessarily. *Primitive* is a relative term used to describe the condition of the most recent common ancestor of a given group. Five digits on the terminal part of the forelimb is the primitive condition for mammals because the common ancestor of all mammals, whatever it was, must have had five digits. We presume this must be so because all, most, or some of the constituent taxa of most orders of mammals have five digits on that part; those mammals that have fewer digits frequently show signs of having had more in their early embryonic stages. Finally, reptiles and amphibia, close relatives of mammals, also typically have five digits; those with fewer, likewise seem to be later developments. In this respect, human hands are primitive, but they are also specialized by virtue of long grasping fingers and an opposable thumb.

If after a given group first appears, it undergoes a significant amount of evolutionary diversification, the greater part of this change inevitably involves invention and elaboration of structure and behavior, not to improve on the ancestral adaptation to the original niches, but rather to exploit new environmental possibilities or invade new niches: in short, to specialize. Most adaptive radiations or truly large-scale examples of diversifications of a single evolutionary stock, as well as the more modest bursts of organic diversity, involve the development of side branches of stocks that are more anatomically and ecologically specialized. Thus, the relatively more generalized members of a group are usually, although not invariably, those that retain the primitive unelaborated anatomy and behavior. That is why most long-lived structural types tend to be relatively primitive, and why many primitive types, relative to other members of their own groups, are ecological generalists and thus more likely to enjoy a somewhat higher probability of survival.

Some bona fide living fossils—*Sphenodon* and the clam *Neotrigonia,* for instance, which lack close modern relatives—belong to groups that were considerably more diverse in remote times. But there are other living fossils—*Limulus* and the mollusk *Neopilina* being two—that belong to groups that never attained any appreciable diversity. In these instances, we are not dealing with extinctions that weeded out relatives but rather with the nonevolution of many relatives in the first place. Particularly in the instance of the horseshoe crab, relatively little was accomplished in terms of developing new and different plans of body organization. So a final question is, why? Why didn't the horseshoe crab move from its coastal habitat into another environment and then specialize and thus diversify?

This kind of question—why something did *not* evolve—is both intriguing and difficult to answer. Here again, the notion of competition seems to offer a clue. Hindsight indicates that the general body plan of true crabs proved more adaptable and more modifiable than that of the horseshoe crab, thus allowing specialization in such all-important functions as feeding and locomotion. True crustaceans, not horseshoe crabs, were accordingly able to exploit microhabitats, and they therefore diversified far more than the horseshoe crab ever could. As I said earlier, *Limulus* walks, burrows, and swims pretty well, and no true crab known to me can do all three as well. But *Limulus* cannot burrow as efficiently as the mole crab *Emerita* or walk or swim as well as some other crab species can. Without the competitive presence of the crustaceans, perhaps the horseshoe crab would have specialized and diversified more than it did. But if it had specialized, it might not have survived as long as it has.

PART 6
EXTINCTION

Extinction seems the very antithesis of evolution. I used to avoid the subject as much as possible for that very reason: preferring to focus on the *positive* side, looking at the causes of evolutionary change, I thought extinction merely recorded life's evolutionary failures, a phenomenon of little significance when it comes to understanding the origin of new structures and entire groups of organisms.

I could not have been more wrong. Like the phenomenon of living fossils, where analysis of evolutionary stability sheds light on the hows and whys of evolutionary change, extinction, the failure to persist, reveals much about why organisms, and entire taxa, do persist. Remember, evolution in its broadest sense is concerned with maintenance and transmission of genetically based information—and not solely the modification of that information through time. Extinction is the failure to maintain and transmit that information.

Extinction plays more than a strictly heuristic role in helping us to understand the ins and outs of the evolutionary process. In periodically removing species and larger-scale taxa from the ecological arena, extinction opens up vacancies that sooner or later (and it is usually sooner) will be filled—either by organisms still alive and well and living elsewhere who are able to invade and colonize an area, or by the evolutionary production of the truly new. It is now my conviction that without extinction, little in the way of evolution would ever have occurred on planet earth.

Paleontologists often distinguish between "background extinction"—a statistically nearly constant rate of disappearance of species—and "mass extinction," which cuts right across lines of genealogical affinity, affecting many unrelated species nearly simultaneously. In the early nineteenth century, years before Darwin published his epochal book in 1859, geologists were already busy recognizing subdivisions of geological time, based on the natural breaks in distribution of fossils in the rocks they were studying. The Paleozoic, Mesozoic, and Cenozoic Eras formed the three grand subdivisions of what has turned out to be the past 570 million years. There seemed to be rather marked differences in the gross characteristics of these divisions of ancient, middle, and modern forms of life. And we now know that the divisions were sparked by two of the most profound mass extinction events in the entire history of complex life. Lesser extinction events mark the more minor subdivisions, the periods and epochs.

The causes of extinction events have long provoked debate among biologists and paleontologists. The following essays reflect much of the range of thinking on the causes of extinctions—though all were written rather recently, and tend to see extinctions mostly as a function of ecosystem collapse, brought on for the most part by changes in the physical environment. Competition between groups, mentioned in several of the pieces, has lost some of its former popularity—all to the good, given the current status of the concept of interspecific competition and the difficulty (as Michael Benton points out in his essay) of applying the concept to larger-scale taxa.

In the first essay, Stephen Jay Gould (in the first piece he ever wrote for *Natural History,* and his only noncolumn article) examines the case of the "Irish Elk." This enormous Ice Age deer had been the premier example of "orthogenesis"—the idea that evolution is directional, and propelled by factors intrinsic to the biology of organisms themselves. Obviously, or so the story went, such inadaptive structures as those grotesque antlers could not have been fashioned by natural selection, particularly since the antlers were held to have been the culprits that caused the demise of that species of giant deer. Gould recounts the Darwinian rejoinder—but then opts for an environmental explanation for the extinction, reminding us that the Irish Elk was but one of many faunal elements that disappeared as the Pleistocene drew to a close, and as environments changed radically in Europe.

Still with the Ice Age, the essay by Hazel and Paul Delcourt takes us to Pleistocene North America, examining the botanical side of the ledger. It is clear that, as environments change, there are three options open to organisms. Standard Darwinian theory stresses adaptive evolutionary modification in response to such changing conditions—provided, of course, that the necessary genetic variation exists to allow change to occur. Extinction is the second possibility. Here, the Delcourts examine the third possibility, the *avoidance* of extinction by moving. Most plants, of course, are firmly rooted to the ground, and must die if their habitat is too greatly altered. But plants send propagules out, and their distributional ranges are altered through air, water, and animal-aided dispersal of seeds. Thus plants can move, and can "track" suitable habitats which move around according to the nature of the environmental change. In the Pleistocene, that change was montane and continental ice sheets. Habitats, in general, shifted southward in response.

The most famous extinction event of all, at the end of the Cretaceous Period, when the dinosaurs and much of the rest of life underwent a dramatic cutback, provides the focus for Neil Landman's analysis of the demise of the marine ammonites. Concomitantly, Landman wonders why, as the ammonites went, their first cousins, the nautilids, managed to make it through. Indeed, the modern pearly nautilus, still very much a denizen of the Indo-Pacific, is a classic "living fossil." Landman offers us a two-pronged theory to account for the different histories of these two groups of shelled cephalopods. First, nautilids seem to be ecologically rather generalized, and Landman marshals several inferences about ammonite paleobiology implying that many ammonite species were ecologically rather specialized. Generalists may be supposed to have a higher probability of survival than specialists—an argument familiar from the preceding essays on living fossils. But Landman also notes the larger context of these events. The Cretaceous extinction, now widely (though I think erroneously) considered the effect of a collision between the earth and either an asteroid or a comet, in any case involved a host of only remotely related species. It was an ecosystem collapse, and Landman notes that ammonites may well have been part of the plankton community of the sea, a community known to have been devastated as the algae at the very base of the food chain suffered mass extinction—perhaps engendered by that extraterrestrial impact.

In the final essay of part 6, Michael Benton develops an intriguing variant of a currently popular theme linking extinction with evolution. Mammals, on earth at least as far back as Triassic times, remained the "rats of the Mesozoic" living in the "interstices of the dinosaurs' world" (to combine phrases of Alfred S. Romer and Stephen J. Gould). The moral: we and most of our mammalian kin would probably not have evolved if extinction hadn't knocked the dinosaurs out of the great array of terrestrial niches they had occupied throughout much of the Mesozoic. Benton shows us that much the same can be said for the dinosaurs' rise to dominance as is so often said about mammals: in particular, Benton thinks the dinosaurs were just plain lucky that the herbivorous rhynchosaurs, who were enjoying ever increasing success as the Triassic wore on, suffered their own mass extinction—probably, he thinks, because the climate was drying up and their favorite food plant itself became extinct. Benton's essay helps us to understand just how extinction has reset the evolutionary clock—and that the process has gone on many times, and not only at the end of the Cretaceous.

21

The Misnamed, Mistreated, and Misunderstood Irish Elk

STEPHEN JAY GOULD

"Nature herself seems by the vast magnitude and stately horns she has given this creature, to have singled it out as it were, and showed it such regard, with a design to distinguish it remarkably from the common herd of all other smaller quadrupeds."

Thomas Molyneux, 1697

The Irish elk, the Holy Roman Empire, and the English horn form a strange ensemble indeed. But they do share a common distinction—they have completely inappropriate names. The Holy Roman Empire, Voltaire tells us, was neither holy, nor Roman, nor an empire. The Irish elk was neither exclusively Irish, nor an elk. But it was the largest deer that ever lived. Its antlers were enormous; if we enlarge any other deer to the size of an Irish elk without changing its shape, its antlers will still fall far short of the "spacious horns" of the giant deer that so astounded Molyneux in 1697.

Although the *Guinness Book of World Records*—ignoring fossils—honors the American moose, the antlers of the Irish elk have never been exceeded or even approached in the history of life. Reliable estimates of their total span range up to 12 feet. This figure is all the more impressive when we recognize that, as in all other true deer, the antlers were probably shed and regrown annually.

Fossil antlers of the giant deer have long been known in Ireland, where they occur in lake sediments underneath peat deposits. Before attracting the attention of scientists, they have been used as gateposts and, in one instance, even as a temporary bridge to span a rivulet in County Tyrone. One story, probably apocryphal, tells of a huge bonfire made of giant deer bones and antlers in County

Antrim to celebrate the victory over Napoleon at Waterloo. They were called elk because the European moose (an "elk" to Englishmen) was the only familiar animal with antlers that even approached those of the giant deer in size.

The first known drawing of giant deer antlers dates from 1588. Nearly a century later, Charles II, who reigned from 1660-1685, received a pair of antlers and, according to Dr. Molyneux, "valued them so highly for their prodigious largeness" that he set them up in the horn gallery of Hampton Court, where they "so vastly exceed" all others in size "that the rest appear to lose much of their curiosity."

Ireland's exclusive claim vanished in 1746 (although the name stuck) when a skull and antlers were unearthed in Yorkshire, England. The first continental discovery followed in 1781 from Germany, while the first complete skeleton, (still standing in the museum of Edinburgh University) was exhumed from the Isle of Man in the 1820s.

We now know that the giant deer ranged as far east as Siberia and China and as far south as northern Africa. Specimens from England and Eurasia are almost always fragmentary, and nearly all the fine specimens that adorn so many museums throughout the world are from Ireland. The giant deer evolved during the glacial period of the last few million years and may have survived to historic times in continental Europe, but it became extinct in Ireland about 11,000 years ago.

"Among the fossils of the British empire," wrote James Parkinson in 1811, "none are more calculated to excite astonishment." And so it has been throughout the history of paleontology. Putting aside both the curious anecdotes and the sheer wonder that immensity always inspires, the importance of the giant deer lies in its contribution to debates about evolutionary theory. Every great evolutionist has used the giant deer to defend his favored views. The controversy has centered around two main issues: (1) Could antlers of such bulk be of any use? and (2) Why did the giant deer become extinct?

Since debate has long centered on the reasons for the Irish elk's extinction, it is ironic that the primary purpose of Molyneux's original article was to argue that the deer must still be alive. Many seventeenth-century scientists maintained that the extinction of any species would be inconsistent with God's goodness and perfection. Dr. Molyneux's 1697 article, the first scientific description of the giant deer, begins: "That no real species of living creatures is so utterly extinct, as to be lost entirely out of the World, since it was first created, is the opinion of many naturalists; and 'tis grounded on so good a principle of Providence taking care in general of all its animal productions, that it deserves our assent."

Yet the giant deer no longer inhabited Ireland, and Molyneux was forced to search elsewhere. After reading travelers' reports of the American moose's antler size, he concluded that the Irish elk was the same animal; the tendency toward exaggeration in such accounts is apparently universal and timeless. Since he could not find either a figure or an accurate description of the moose, his conclusions are not as absurd as modern knowledge would indicate. Molyneux attributed the giant deer's demise in Ireland to an "epidemick distemper," caused by "a certain ill constitution of air."

For the next century arguments followed along Molyneux's line—to which modern species did the giant deer belong? Opinion was equally divided between the moose and the reindeer.

As eighteenth-century geologists unraveled the fossil record of ancient life, it became more and more difficult to argue that the odd and unknown creatures revealed by fossils were all still living in some remote portions of the globe. Perhaps God had not created just once and for all time; perhaps He had experimented continually in both creation and destruction. If so, the world was surely older than the 6000 years that literalists allowed.

The question of extinction was the first great battleground of modern paleontology. In America, Thomas Jefferson maintained the old view, while Georges Cuvier, the great French paleontologist, was using the Irish elk to prove that extinction did occur. By 1812 Cuvier had resolved two pressing issues: by using minute anatomical descriptions, he proved that the Irish elk was not like any modern animal; and by placing it among many fossil mammals with no modern counterparts, he established the fact of extinction and set the basis for the geologic time scale.

Once the fact of extinction had been settled, debate moved to the time of the event: in particular, had the Irish elk survived the flood? This was no idle matter, for if the flood or some previous catastrophe had wiped out the giant deer, then its demise had natural (or supernatural) causes. Archdeacon Maunsell, a dedicated amateur, wrote in 1826: "I apprehended they must have been destroyed by some overwhelming deluge." A certain Dr. MacCulloch even believed that the fossils were found standing erect, noses elevated—the deer's final gesture to the rising flood, as well as their final plea: don't make waves.

If, however, they had survived the flood, then their exterminating angel could only have been the naked ape himself. Gideon Mantell, writing in 1851, blamed Celtic tribes; Hibbert implicated the Romans and the extravagant slaughters of their public games. Lest we assume that man's destructive potential was recognized only recently, Hibbert wrote in 1830: "Sir Thomas Molyneux conceived that a sort of distemper, of pestilential murrain, might have cut off the Irish elks.. . .It is, however, questionable, if the human race has not occasionally proved as formidable as a pestilence in exterminating from various districts, whole races of wild animals."

In 1846 Britain's greatest paleontologist, Sir Richard Owen, reviewed the evidence and concluded that the giant deer, in Ireland at least, had perished before man's arrival. By this time, Noah's flood as a serious geologic proposition had passed from the scene. What then had wiped out the giant deer?

Within ten years of Charles Darwin's publication of the *Origin of Species* in 1859, virtually all scientists had accepted the *fact* of evolution. But the debate about causes and mechanisms was not resolved (in Darwin's favor) until the 1930s. Darwin's theory of natural selection requires that all evolutionary changes be adaptive—that is, that they be useful to the organism. Therefore, the anti-Darwinians searched the fossil record for cases of evolution that could not have benefited the animals involved.

The theory of orthogenesis became the touchstone for anti-Darwinian paleontologists, for it claimed that evolution proceeded in straight lines, which natural selection could not regulate. Certain trends, once started, could not be stopped even if they led to extinction. Thus, it was said that certain oysters coiled their valves upon each other until they sealed the animal permanently within; that saber-toothed tigers could not stop growing their teeth or mammoths their tusks.

But by far the most famous example of orthogenesis was the Irish elk. The giant deer had evolved from small forms with even smaller antlers. Although the antlers were useful at first, their growth could not be contained and, like the sorcerer's apprentice, the giant deer discovered only too late that even good things have their limits. Bowed by the weight of their cranial excrescences, caught in the trees or mired in the ponds, they died. What wiped out the Irish elk? They themselves or, rather, their own antlers did.

In 1925 the American paleontologist R. S. Lull invoked the giant deer to attack Darwinism: "Natural selection will not account for overspecialization, for it is manifest that, while an organ can be brought to the point of perfection by selection, it would never be carried to a condition where it is an actual menace to survival. . .[as in] the great branching antlers of the extinct Irish deer."

Darwinians, led by Julian Huxley, led a counterattack in the 1930s. Huxley noted that as deer get larger—either during their own growth or in the comparison of related adults of different sizes—the antlers do not increase in the same proportion as body size; they increase faster, so that the antlers of large deer are not only absolutely larger but also relatively larger than those of small deer. For such regular and orderly change of shape with increasing size, Huxley used the term allometry.

Allometry provided a comfortable explanation for the giant deer's antlers. Since the Irish elk had the largest body size of any deer, its relatively enormous antlers could have been the simple consequence of an allometric relationship present among all deer. We need only assume that increased body size was favored by natural selection; the large antlers might have been an automatic consequence. They might even have been slightly harmful in themselves, but this disadvantage was more than compensated for by the benefits of larger size, and the trend continued. Of course, when problems of larger antlers outweighed the advantages of larger bodies, the trend would cease since it could no longer be favored by natural selection.

Almost every modern textbook of evolution presents the Irish elk as a model case in this light, citing the allometric explanation to counter orthogenetic theories. As a trusting student, I had assumed that such constant repetition was firmly based on copious data. Later I discovered that textbook dogma is self-perpetuating; therefore, three years ago I was disappointed, but not really surprised, to discover that this widely touted explanation was based on no data whatsoever. Aside from a few desultory attempts to find the largest set of antlers, no one had ever measured an Irish elk. Yardstick in hand, I resolved to rectify this situation.

The National Museum of Ireland in Dublin has seventeen specimens on display and many more, piled antler upon antler, in a nearby warehouse. Most large museums in Western Europe and America own an Irish elk, and the giant deer adorns many trophy rooms of English and Irish gentry. The largest antlers grace the entranceway to Adare Manor, home of the Earl of Dunraven, in Ireland. The sorriest skeleton, also in Ireland, sits in the cellar of Bunratty Castle, where many merry and slightly inebriated tourists repair for coffee each evening after a medieval banquet. This poor fellow, when I met him early the morning after, was smoking a cigar, missing two teeth, and on the tines of his antlers, he carried three coffee cups. For those who enjoy invidious comparisons, the largest antlers in America are at Yale; the smallest in the world at Harvard.

To determine if the giant deer's antlers increased allometrically, I compared antler and body size. For antler size, I used a compounded measure of antler length, antler width, and the lengths of major tines. Body length, or the length and width of major bones, might be the most appropriate measure of body size, but I could not use it because the vast majority of specimens consist of only a skull and its attached antlers. Moreover, the few complete skeletons are invariably made up of several animals, much plaster, and, occasionally, ersatz (the first skeleton in Edinburgh once sported a horse's pelvis). Skull length therefore served as my measure of overall size. This is not so bad a resolution as it may at first appear. Body length and weight depend strongly upon age and condition; these are extraneous factors with strong effects upon antlers that can easily confuse a primary correlation with size (measures for antler size can be corrected for age). The skull, however, reaches its final length at a very early age (all my specimens are older) and does not vary thereafter; it is, therefore, a good indicator of body size. My sample included 79 skulls and antlers from museums and homes in Ireland, Britain, continental Europe, and the United States.

My measurements showed a strong position correlation between antler size and body size, with the antlers increasing in size two and one-half times faster than body size from small to large males. This is not a plot of individual growth; it is a representation for adults of different body size. Thus, the allometric hypothesis is affirmed. If natural selection favored large deer, then relatively larger antlers would appear as a correlated result of no necessary significance in itself.

Yet, even as I affirmed the allometric relationship, I began to doubt the traditional explanation—for it contained a curious remnant of the older, orthogenetic view. It assumed that the antlers are not adaptive in themselves and were tolerated only because the advantages of increased body size were so great. But why must we assume that the immense antlers had no primary function? Equally possible is the opposite interpretation: that selection operated primarily to increase antler size, thus yielding increased body size as a secondary consequence. The case for inadaptive antlers has never rested on more than the subjective wonderment born of their immensity.

Views long abandoned often continue to exert their influence in subtle ways. The orthogenetic argument lived on in the allometric context proposed to replace it. I believe that the supposed problem of "unwieldly" or "cumbersome" antlers is an illusion rooted in a notion now abandoned by students of animal behavior.

To nineteenth-century Darwinians, the natural world was a cruel place. While modern scientists tend to measure the benefits of evolution in terms of successful reproduction, last century's Darwinians assessed evolution in terms of battles won and enemies destroyed. In this context, antlers were viewed as formidable weapons to be used against predators and rival males. In his *Descent of Man* (1871), Darwin toyed with another idea: that antlers might have evolved as ornaments to attract females. "If, then, the horns, like the splendid accouterments of the knights of old, add to the noble appearance of stags and antelopes, they may have been modified partly for this purpose." Yet he quickly added that he had "no evidence in favor of this belief," and went on to interpret antlers according to the "law of battle" and their advantages in "reiterated deadly contests." All early writers assumed that the Irish elk used its antlers to kill wolves and drive off rival males in fierce battle. To my knowledge this view has been challenged only by the Russian paleontologist L. S. Davitashvili, who in 1961 asserted that the antlers functioned primarily as courtship signals to females.

Now, if antlers are weapons, the orthogenetic argument is appealing, although I must admit that ninety pounds of broad-palmed antler, regrown annually and spanning twelve feet from tip to tip, seems even more inflated than our current military budget. Therefore, to preserve a Darwinian explanation, we must invoke the allometric hypothesis in its original form.

But what if antlers do not function primarily as weapons? Modern studies of animal behavior have generated an exciting concept of great importance to evolutionary biology: many structures previously judged as actual weapons or devices for display to females are actually used for ritualized combat among males. Their function is to prevent actual battle (with its consequent injuries and loss of life) by establishing hierarchies of dominance that males can easily recognize and obey.

Antlers and horns are a primary example of structures used for ritualized behavior. They serve, according to Valerius Geist, as "visual dominance-rank symbols." Large antlers confer high status and access to females. Since there can be no evolutionary advantage more potent than a guarantee of successful reproduction, selective pressures for large antlers must often be intense. As more and more horned animals are observed in their natural environment, older ideas of deadly battle are yielding to evidence of fighting in ways clearly designed to prevent bodily injury or of purely ritualized assertion of dominance without body contact. This has been observed in red deer by Beninde and Darling, caribou by Kelsall, and in mountain sheep by Geist.

As devices for display among males, the enormous antlers of the Irish elk finally make sense as structures adaptive in themselves. Moreover, as R. Coope of Birmingham University pointed out to me, the detailed morphology of the antlers can be explained, for the first time, in this context. Deer with broad-palmed antlers tend to show the full width of their antlers in display. The modern fallow deer (considered by many as the Irish elk's nearest living relative) must rotate its head from side to side in order to show its palm. This would have created great problems for giant deer, since the torque produced by swinging ninety-pound antlers would have been immense. But the antlers of the Irish elk were arranged so ingeniously that it only had to bow its head once to display all facets of its antlers. When the deer was looking straight ahead, the palm was fully displayed. With its head down, both the length of the giant deer's antlers and the strength of their tines were strikingly evident. (In any case the tines could hardly have been effective in battle since they pointed backwards and could not be directed toward a facing enemy unless the deer held its head down between its legs.) Therefore, both the unusual configuration and the enormous size of the antlers can be explained by postulating that they were used for display rather than for combat.

If the antlers were adaptive, why did the Irish elk become extinct (at least in Ireland)? The probable answer to this old dilemma is, I am afraid, rather commonplace. The giant deer flourished in Ireland for only the briefest of times— during the so-called Allerod interstadial phase at the end of the last glaciation. This period, a minor warm phase between two colder epochs, lasted for about 1,000 years, from 12,000 to 11,000 years before the present. (The Irish elk had migrated to Ireland during the previous glacial phase when lower sea levels established a connection between Ireland and continental Europe.) While it was well adapted to the grassy, sparsely wooded, open country of Allerod times, it apparently could not adapt either to the subarctic tundra that followed in the next cold epoch or to the heavy forestation that developed after the final retreat of the ice sheet.

Extinction is the fate of most species, usually because they fail to adapt rapidly enough to changing conditions of climate or competition. Darwinian evolution decrees that no animal shall actively develop a harmful structure, but it offers no guarantee that useful structures will continue to be adaptive in changed circumstances. The Irish elk was probably a victim of its own previous success.

22

Ice Age Haven for Hardwoods

HAZEL R. DELCOURT
and
PAUL A. DELCOURT

At the peak of the last continental glaciation, 18,000 years ago, eastern North America bore little resemblance to the landscape encountered by European settlers in the seventeenth century. Today, we tend to assume that for untold thousands of years before the clearing of the land, extensive regions of the continent were dominated by cool-temperate, mesic (moisture loving) deciduous trees—beech, sugar maple, yellow birch, tulip tree, basswood, walnut, and certain oaks and hickories. This assumption is justified only to a certain degree because these trees have not been in the northeastern and midwestern portions of North America continuously.

During previous warm interglacial periods, the temperate deciduous forests were probably distributed very much as they are today. But during glacial times, such as the last major period of glaciation, which lasted from 80,000 to as recently as 10,000 years ago, climatic conditions would have been intolerable for this kind of hardwood forest. For decades, plant geographers were frustrated in their efforts to discover just where populations of these now widespread tree species might have found refuge during glacial times. Only in the last decade have clear answers begun to emerge.

During peak, or full, glacial times, sea level was approximately 300 feet lower than it is today, and great quantities of fresh water were locked up in extensive continental ice sheets. Current evidence indicates that the climatic changes that led to the great ice sheets had an enormous influence on the face of eastern North America. The Laurentide Ice Sheet extended southward nearly to the confluence of the Ohio and Mississippi rivers, and in certain areas, tundra bordered its southern edge. Boreal coniferous forest occupied a latitudinal belt south of the tundra from Iowa to Pennsylvania, southward to Arkansas, northern Mississippi and Alabama, and across the Carolinas. Jack pine dominated in the east, and spruce was abundant to the west across the Great Plains. In the alluvial valley of the Mississippi River, white spruce and tamarack extended southward to Louisiana, occupying the extensive flats of braided streams that carried glacial meltwater and sediment to the Gulf of Mexico.

At this time, a steep climatic gradient ran from northern Mississippi across central Alabama and Georgia to southern South Carolina, separating boreal from warm-temperate vegetation. This gradient (the Polar Frontal Zone) was steeper 18,000 years ago than it is today. Then, the yearly patterns of atmospheric circulation were usually marked by a persistent west-east flow, rather than the seasonal north-south shifts in the jet stream and associated air masses that characterize interglacial climates. The climatic boundary separating boreal from warm-temperature ecosystems penetrated far into the southeastern United States during glacial times. South of this limit, a forest of oaks, hickories, and southern pines occupied much of the Gulf and southern Atlantic coastal plains. Sand dune scrub vegetation covered most of the central and southern Florida peninsula, where the lowered sea level drew down the local water table by at least 65 feet.

Because of the extensive climatic changes and because few areas between the Appalachians and the Atlantic had soils rich enough to support mesic hardwood communities, habitats available for colonization by cool-temperate hardwood species were without question very restricted during the last glacial maximum. Biogeographer and paleoecologist Edward S. Deevey theorized as early as 1949 that these plant species must have been displaced far south of their present ranges, into southern Florida and even Mexico. Evidence to support this view was scant and controversial, however, for at that time, few full glacial sites had been identified in the region south of the glacial margin in eastern North America.

A more conservative view was held by the prominent forest ecologist E. Lucy Braun, who argued that the present-day forests of the Cumberland Mountains of eastern Kentucky, as well as certain hardwood forests of the southern Appalachian Mountains, represent relicts of plant communities established there at least twenty-four million years ago, during the Tertiary period. This belief was based upon the co-occurrence in these areas today (and the assumed co-occurrence in the past) of mesic hardwood trees with numerous species of forest-understory shrubs and herbaceous plants that have taxonomic affinities to fossil plants known from the Tertiary. Many of these understory species are narrowly endemic, that is, found only within a restricted portion of the ranges of mesic deciduous tree species and particularly within the southern Appalachian Mountains. How, Braun asked, could such plant species—known to have limited dispersal mechanisms and competitive abilities—ever have survived if encroached upon by boreal species? Her answer was that the endemic plants could not have survived dramatic climatic cooling, but rather that they persisted within intact hardwood communities throughout the Pleistocene epoch (2.5 million years ago to 12,000 years ago) of the Quaternary period. Braun contended that there has been no vegetational change in the unglaciated Southeast.

Resolution of the debate between Deevey and Braun had to wait until the discovery of fossil sites showing just how far plant species had been displaced during episodes of climatic cooling in the Pleistocene. By the 1970s paleoecologists analyzing pollen preserved in lake sediments along the central Atlantic seaboard had conclusively demonstrated that full glacial forests in the Carolinas were boreal in character, dominated by jack pine trees. Farther south, Bill Watts, of Trinity College in Ireland, carried out an extensive search for full glacial records of deciduous forest species on a transect across Georgia into Florida. Among his significant findings were several ponds in northwestern Georgia with basal sediments containing jack pine needles; the pollen record of these sediments were also dominated by jack pine. The southernmost site on the transect, Lake Annie, revealed that the central Florida peninsula was covered by sand dunes and dry, scrubby vegetation throughout the last glacial period.

So the question persisted: where were the full glacial refuges for the deciduous trees? If we concentrated on life history characteristics of specific trees, would that help us know where to look? Beech, for example, is relatively slow growing, requires rich soil, and is intolerant of drought. Its seeds are commonly dispersed by small mammals and by birds such as the bluejay and, formerly, the passenger pigeon. Would the beech's chances of escaping to sites south of the ice sheet have been greater along major flyways such as the Mississippi Valley? Were heavy-seeded species, such as the walnut and the Kentucky coffee tree, dependent on now-extinct large mammals, such as the Pleistocene horse and mastodon, for dispersal to refuge areas? One of the factors that define the northernmost limits of many of today's deciduous species, such as sugar maple, oaks, hickories, ash, and elm, is the severity of winter temperatures. During the climatic cooling that led to the onset of glaciation, was the northernmost range of these trees progressively truncated as the mean winter position of the Polar Frontal Zone moved farther and farther to the south?

During the 1970s, the network of known plant-fossil sites grew, and the search for full glacial refuges intensified. Our own strategy concentrated on looking for localities with fertile soil in regions whose ice age climates may have been moderated by water bodies such as the Gulf of Mexico or the Mississippi River. In 1976, we were excavating sediments associated with the remains of a 17,000-year-old mastodon from a river terrace along Nonconnah Creek, within the city limits of Memphis, Tennessee. To our great excitement, we found abundant twigs, needles, and cones of white spruce, along with walnuts, hickory nuts, acorns, hazelnuts, tulip tree fruits, and beechnuts. Pollen from these sediments also recorded the local presence of populations of sugar maple, birch, elm, and ash. The Nonconnah site thus provided us with new data and important clues toward solving the mystery of the full glacial habitats of these forest species.

Our explanation for the co-occurrence of spruce and deciduous trees at Nonconnah Creek, as well as at other late glacial localities now known from as far south as southeastern Louisiana, takes into account the special habitats provided by a unique natural region. Known as the Blufflands, this region is a belt of hilly land extending along the eastern wall of the Mississippi River Valley from southern Illinois to southern Louisiana. The sedimentary deposits that form the Blufflands are composed of thick accumulations of loess—silt that was swept by wind from the braided stream surfaces of the Mississippi River Valley during the late Pleistocene and deposited eastward as a mantle covering the uplands in a narrow zone. As this fertile, soft substrate eroded, steep ravines formed, creating a cool microclimate that has persisted into the modern era. As a result, an unusual assemblage of cool-temperate plants thrives as far south as the Tunica Hills of southeastern Louisiana.

We hypothesize that in the Pleistocene, as today, the loess deposits would have provided excellent conditions for the growth of mesic hardwood communities. During the last full glacial period, cold glacial meltwater funneling down the alluvial valley would have had a cooling influence upon adjacent uplands. Contact of the cold river water with relatively warm, moist air from the Gulf of Mexico would have given rise to persistent fogs along the Blufflands. Frequent fogs would have supplied moisture, increased cloud clover, and provided a cooler, more humid microclimate extending south of the Polar Frontal Zone into the Gulf Coastal Plain. These factors would have allowed populations of boreal tree species to expand southward along the alluvial valley of the Mississippi River, and they would have also permitted mesic deciduous tree species to persist within the adjacent uplands of the Blufflands. Similar full glacial refuges for deciduous trees may have occurred along other major river systems of the southeastern United States. Recently, Bill Watts has obtained paleoecological evidence from Sheelar Lake, a small pond in northern Florida, that indicates that pockets of mesic deciduous forest may also have persisted there in deeply dissected ravines adjacent to spring-fed sinkholes.

The importance of north-south corridors for the survival of deciduous forest can be seen clearly by comparing the situation in North America with that in Europe. Most of Europe's deciduous forest species, known from fossil remains of the Tertiary age, died out during the past 2.5 million years. The extinctions were progressive, with more species being lost with each of the approximately twenty cycles of advance and retreat of continental glaciers. During the height of each glacial period, cold, dry air masses penetrated far south of the continental ice sheets onto the European continent. At such times, suitable refuge areas for deciduous forest trees existed along the Mediterranean coast, but the east-west trending mountain ranges of the Alps and Pyrenees formed a barrier to plant migration that few species could pass.

The relative continuity of suitable habitats in eastern North America would also have been crucial to the survival of deciduous species in late glacial and early interglacial times. It would have made possible the rapid dispersal and establishment of these trees as they advanced northward with the warming climate. Unlike the onset of continental glaciation, which involves gradual cooling over as much as 90,000 years, interglacial climatic warming is relatively rapid. It occurs within several thousand years and dramatically affects the distribution of tree species.

The last major interval of climatic warming and deglaciation began about 16,500 years ago. As the glaciers retreated, vast plains of glacial drift material— clay, sand, gravel, and boulders—blanketed most of the midwestern and northeastern portions of North America. Embedded in these newly deglaciated landscapes were remnant ice blocks. The lakes that formed as these ice blocks melted have since accumulated sediment and plant fossils, giving us a wealth of direct evidence about the most recent northward movements of trees. (Little concrete information is available for *southward* migration of trees during times of glaciation because each time the ice advanced, it obliterated lakes formed by the previous glacial retreat.)

Tree species migrate at different rates and in different directions according to their particular strategies for seed dispersal, habitat requirements, and chance historical events. During the last full glacial period, for example, jack pine was widespread across eastern North America and extended as far south as northwestern Georgia and central South Carolina. This species composed as much as 80 percent of the forests across the region until 16,500 years ago. Then, as the climate warmed up again, spruce and temperate hardwood species proved to be better competitors. Jack pine seedlings do not fare well in the dense shade of hardwood trees, and the species died out rapidly along the southern margin of the boreal forest region, from Missouri to North Carolina.

Like many other boreal species, jack pine did migrate northward in the path of the retreating Laurentide Ice Sheet, but the movement was limited by the slow rate at which the ice melted. Temperate, hardwood deciduous trees were never far behind in the northward march, and jack pine, an early successional species less suited to the rapidly changing climatic conditions, was never able to establish itself for more than a relatively short time in any one area. Today, jack pine forests prevail from Minnesota to Michigan, on sandy plains of glacial outwash where frequent fires and nutrient-poor soils eliminate potential competitors. They also occupy extensive portions of Manitoba and central Ontario where they are beyond the northern limit of most deciduous trees.

The migrational histories of many deciduous trees were unlike that of jack pine. One of the slowest to colonize the deglaciated north was the American chestnut. The refuge areas for this species during the full glacial period were in southern Alabama and probably elsewhere along the Gulf Coastal Plain. Chestnut responded slowly to climatic change, appearing in the central Appalachian Mountains only about 5000 years ago and reaching its ultimate northward limit in northern New England as recently as 2000 years ago. Individuals of this species are relatively long-lived (up to 300 years) and attain optimal growth on dry to moist sites at middle and high elevations of the Appalachian Mountains. Chestnut has large heavy fruits that are dispersed by mammals, which may explain the slow rate of this species' postglacial migration. (Unfortunately, after having taken nearly 15,000 years to reach its broadest distribution and greatest dominance, chestnut was virtually eliminated from the eastern deciduous forests in less than forty years. A fungal blight was inadvertently introduced to New York in 1904, and by 1940 the chestnuts were gone.)

The migrational histories of other tree species fall between these two extremes. By 12,000 years ago beech had expanded its range broadly across the southeastern United States, and the species subsequently migrated northward along and east of the Appalachian Mountains, crossing westward from New York into Ontario about 7000 years ago and reaching its general northern and western limits in Michigan and Wisconsin by 3000 years ago. Maple and elm moved in broad migration fronts from southwest to northeast across the region, averaging between 650 and 800 feet per year, a speed perhaps abetted by wind dispersal of their light, winged fruits. Species that are now characteristic of the northern mixed conifer-hardwood forest of the Great Lakes region achieved their present distributions and relative dominances only in the past 4000 years.

The broad patterns of late Quaternary vegetational history in eastern North America have been established during the past several decades. Nevertheless, current hypotheses are built from fragmentary evidence and will certainly be refined in the future. There are still many tree species for which full glacial refuges are not known. Red pine, eastern white pine, hemlock, and white birch, for example, all appear in late glacial fossil records, but significantly after climatic warming had begun. Questions remain as to whether the diversity of habitats in the southern Appalachian Mountains was sufficient to have provided refuges for these trees. For other, endemic plant species of the southeastern United States, fossil evidence may never be found to document their history, either because of their inherent rarity or because they occur in habitats that are not close to lakes, bogs, or swamps, the environments that are the richest source of fossil pollen and other plant parts. However, the great promise that the unglaciated southeastern United States holds for paleoecologists is that new localities such as Nonconnah Creek will continue to be discovered, leading to important new insights and changing the way in which we view the history of our forests.

23

Not To Be or To Be?

NEIL H. LANDMAN

Toward the end of the Cretaceous period, about 65 million years ago, a veritable who's who of the ancient animal kingdom—from giant land-based dinosaurs to various microscopic marine plankton—became extinct. Explanations for these dramatic disappearances range from catastrophic events, such as collisions with asteroids or streams of comets, multiple simultaneous volcanic eruptions, or massive freshwater flooding of the earth's oceans to more gradual events such as a lowering of sea level over several million years, resulting in fewer habitable living areas for marine creatures and a worldwide harshening of climate.

Whatever the causes of the extinctions, many animals did survive, even though closely related groups may not have. The dinosaurs and many other reptiles died out, but alligators survived. Among the smaller life forms exhibiting this "differential survival" were the ammonites and the nautilids, related invertebrate mollusks. The ammonites, much more numerous than the nautilids throughout their combined history, became extinct; the nautilids survived. Why?

These two groups of marine animals belong to the Cephalopoda, the most advanced class of mollusks, which also includes squids, cuttlefish, and octopuses. Ammonites and nautilids shared the same general habitat and the bodies of both were encased in chambered calcareous shells. The soft-bodied animal inside the shell lived in the largest chamber, which was open to the outside. Running through the chambers was a small tube (the siphuncle) that permitted the transfer and removal of liquid to control buoyancy. In ammonites, the calcareous partitions (septa) between the chambers were convoluted and folded into marvelously complex patterns. In contrast, the partitions of nautilids were simple curved surfaces, without frilling. These differences in the shapes of the partitions are readily apparent on fossil shells. After an ammonite or nautilid died on the sea bottom, its empty shell and chambers filled with mud, which subsequently turned into rock, leaving an internal mold of the shell. When the outer shell wall dissolved, the edges of the internal partitions were revealed, forming what are known as suture patterns. These suture patterns are used to classify ammonites and nautilids into their respective families.

Ammonites and nautilids also differed in the relative position of their buoyancy tubes and in overall shape. In nautilids, the tube ran through the center of the partitions; in almost all ammonites it ran through the base of the partitions at the margin of the outer shell. Ammonite shell shapes were diverse and ranged from forms that resembled coiled-up snakes (which is what early collectors mistook them for) to elongate cones. Some ammonite shells were covered with fine lines, while others had large, spiny protuberances. In contrast, the shells of coexisting nautilids were generally globular and lacked conspicuous ornament.

For more than 150 million years before the end of the Cretaceous period, the ammonites were so numerous that many observers of the fossil record call the Mesozoic era (extending from 245 to 65 million years ago and comprising the Triassic, Jurassic, and Cretaceous periods) the age of ammonites. During that time the ammonites completely dominated the nautilids by any measure of abundance: number of genera, species, or individuals. In the late Cretaceous, there were 246 genera of ammonites and only 14 of nautilids. The nautilids, although always present, invariably occurred in fewer numbers. One paleontologist, Karl Waage of Yale University, estimated that in some late Cretaceous sedimentary rock strata in South Dakota the number of individuals belonging to a single ammonite species ranged into the hundreds of millions, but the number of nautilids in the same strata was fewer than 10,000.

The relatively rare nautilids, however, went on to evolve a modest number of divergent forms in the Tertiary period, and the pearly nautilus that inhabits the Indo-Pacific today testifies to their continued presence. What is the preserved record of the ammonites and nautilids in the last few million years of the Cretaceous leading up to this turn of events?

The Cretaceous period is subdivided into a number of time stages. The last of these is the Maastrichtian, named for rock strata in Maastrich, Holland, that were formed between 75 and 65 million years ago. The number of ammonite genera in this interval is lower than in preceding late Cretaceous time stages but still totals thirty-five. In contrast, the number of nautilid genera in this period is ten and is similar to that of other late Cretaceous time stages. Six of these genera are definitely reported from Maastrichtian strata; the other four appear in older and younger strata and therefore were alive during this time interval.

The extinction of ammonites at the very end of the Maastrichtian is recorded in only a handful of sedimentary rock sections in western Europe. In one section, at Zumaya, Spain, only four genera of ammonites occur 30 feet below the Cretaceous-Tertiary boundary. However, at Stevens Klint, Denmark, seven genera of ammonites extend right to the boundary, which is marked by an abundance of embryonic and juvenile shells. In all, eleven genera of ammonites have been recorded at the end of the Maastrichtian stage, although given the scarcity of preserved sedimentary rock sections, other genera may have been alive then but not preserved. In contrast, only one nautilid genus is reported from the end of the Maastrichtian, but at least six other genera must have been alive at that time because they are preserved in both older and younger strata, bringing the total number of nautilid genera in this interval to seven.

The ammonites, therefore, persisted right to the end of the Cretaceous. They may have been reduced in generic number and geographic distribution from preceding time stages but they still exceeded the number of nautilid genera. Nor do their previous abundance and diversity hint at their selectivity extinction. How, then, do we account for this outcome?

One explanation may derive from some very general notions about the survival of animal species under stressful conditions. Large size and broad geographic distribution, for example, are thought to improve a species' chances of survival during an environmental catastrophe in which many habitats are destroyed and competition for resources is intensified. Species survival is also thought to depend, however, on factors unrelated to previous performance. A species that relies on only one food source, for instance, may flourish during calm times but may perish during an environmental catastrophe in which its sole food source disappears. Obviously, a species with broad food preferences may fare better under the same circumstances. Similarly, a species that can live in only one habitat may be in greater peril than a closely related species that can tolerate a variety of habitats. A species with relatively wide food and habitat preference is called an ecologic generalist; one with narrow preferences is an ecologic specialist.

These last arguments may apply to the situation between the ammonites and nautilids. A comparison of range of shell shapes and morphological complexity leads us to suspect that ammonite species were ecologic "specialists" relative to nautilid species. Ammonite species came in numerous shapes and sizes and displayed varying degrees of complexity in their shell partitions. Even the species that made up the eleven genera found at the very top of the Maastrichtian strata (that is, those that lived at the end of the period in question) had various shapes—from streamlined disks to elongate cones—which suggest adaptations to different and possibly specialized modes of life. Streamlined forms may have swum above the sea bottom in shallow water, while other species may have floated near the surface. Ammonite jaws also came in a variety of shapes, possible evidence that different ammonites ate different things.

In contrast, Cretaceous nautilid species displayed only a small range of shell shapes, all basically resembling that of the modern nautilus. Their shell partitions were relatively simple and their one jaw type was almost identical to that present in today's nautilus. Studies of the modern nautilus in its natural habitat by Bruce Saunders of Bryn Mawr College and Peter Ward of the University of California at Davis have shown that the nautilus lives over a broad range of water depths from near the surface to 2000 feet deep, although it is most abundant between depths of about 500 and 1500 feet. It swims just above the bottom and is mainly a scavenger with little food preference. Remote deep-sea cameras set up at appropriate depths have revealed that the nautilus displays a feast-or-famine existence— an adaptation to the unpredictability of available food resources. A nautilid species in the Cretaceous may similarly have lived over a broad range of depths and been unspecialized in its food selection. As generalists in these two respects, then, the nautilids may have had the flexibility to survive in the face of deteriorating and even disastrous environmental conditions.

However, another, much more specific explanation alluded to by the geologist Cesare Emiliani of the University of Miami and Peter Ward, and based on my own research, presents itself. Could the differential extinction of the ammonites at the close of the Cretaceous be related to the simultaneous expiration of some other ill-fated group?

At the end of the Cretaceous, 90 percent of all calcareous plankton became extinct. Among the explanations for the extinction of these surface-water organisms are gradual but dramatic changes in surface-water chemistry and ocean currents, and the darkness resulting from an asteroid collision. Such a collision would have cut off sunlight, thereby inhibiting the photosynthesis necessary for the survival of plant plankton. The death of large portions of the plank plankton would, in turn, have led to the death of many of the animal plankton that fed on them.

Because surface waters lack the buffering capacity of deep water, they are sensitive to environmental changes. The microscopic life forms that live near the surface are therefore more vulnerable to physical disturbances than their deep-water relatives. If ammonites, in contrast to nautilids, were part of this plankton community sometime during their early life history, they might consequently have been annihilated in the general extinction of these surface-water forms.

Unraveling the early life histories of extinct animals is difficult. It is generally done by documenting changes in shape from small to large individuals and then seeing if those differences correspond to growth stages of closely related living forms whose development is known. We can thus begin our analysis with the modern nautilus, but even here, little hard evidence exists. Although the nautilus can be maintained in aquariums, all attempts at breeding it have so far been unsuccessful. Nautilus eggs, however, have been found both in aquariums and in nature; they measure more than one inch in diameter, large even for deep-water invertebrates. The eggs were first described by Arthur Willey, a British naturalist who worked in Indonesia at the turn of the century. Willey was unsuccessful in his efforts to breed the nautilus, but he made an educated guess about the size and appearance of the embryonic nautilus shell. He noticed a constriction on the early whorls of the nautilus shell at a size that corresponded to the diameter of the nautilus egg, and concluded that this feature marked the position on the shell at which hatching occurred. Shell whorls up to this point show no evidence of breakage. Signs of damage appear only after the constriction, an observation consistent with the theory that the shell was protected within an egg case up to that point.

Internally, the point of hatching is marked by a reduction in the spacing of interior partitions, which is coincident with the formation of the constriction on the outer shell. Studies of the geochemical makeup of the nautilus shell have also supported this hatching hypothesis. The chemical composition before and after the constriction differs in subtle but significant ways due to the modifying effects of the egg capsule enclosing the developing embryonic shell.

All this evidence suggests that the nautilus hatches at more than one inch in diameter with about seven buoyancy chambers present. Newly hatched animals have rarely been found, however. Only two have ever been caught alive, and those by accident. This scarcity may reflect the low fecundity of adults, and the probability that the large eggs hatch at water depths of several hundred feet and are well concealed from predators—and scientists. In this environment the newly hatched nautilus begins life as an active swimmer and immediately pursues the deep-water scavenging behavior characteristic of adults.

The reconstructed early life history of Cretaceous nautilids is based directly on our own observations of the modern nautilus. Study of one genus that occurs in the Cretaceous and Tertiary, *Eutrephoceras,* reveals a constriction on its shell similar to that present on the modern nautilus, but it occurs at half an inch in shell diameter, indicating an embryonic size that is still relatively large. Observations of the spacing of internal partitions suggest that the embryonic shells contained about four buoyancy chambers, although finds of preserved embryonic shells are rare. We can probably assume that, like modern nautiluses, individuals of this nautilid genus and similar genera in the late Cretaceous were active swimmers at hatching time and promptly followed an adult mode of life in deep water.

Assumptions about the early life history of ammonites are also based on comparisons with today's nautilus and other extant mollusks such as snails and clams. Prominent changes in the shape of the shell are interpreted as indicating abrupt shifts in life history, as at hatching. Because the early shell whorls of ammonites are very small, scanning electron microscopy is used to examine the shell and its component structures.

Despite the diversity in shape and size of adult ammonite shells, the early whorls of all ammonites are essentially identical. Conspicuous changes in the shape and structure of the early whorls invariably occurred at a diameter of about one millimeter (1/25 of an inch), marked by the presence of a constriction in the shell wall. Up to this constriction, the shell is composed of a single prismatic structure; that is, it is composed of minute, elongate crystals perpendicular to the shell surface. As the shell grew past the constriction, another kind of shell material, called nacre (mother-of-pearl), was introduced. Nacreous crystals are flat-lying hexagonal tablets that are responsible for the iridescence observed in all shells. Below one millimeter, the ammonite shell is also covered with a delicate beadlike ornamentation only visible under high magnification. This ornamentation disappeared abruptly as the shell continued to grow and was replaced by new ornamentation and growth lines. At the same time, the shape of the shell might change dramatically. For example, in cone-shaped ammonites, the conical shape began to develop here.

These abrupt shell changes at a diameter of one millimeter suggest that hatching occurred at this size, less than a tenth the hatching size of contemporaneous nautilids. This hypothesis has been supported by the discovery in late Cretaceous rocks in Montana of thousands of preserved embryonic ammonites. Although this find has not attracted the attention focused on the discovery of newly hatched dinosaurs in younger rock strata a few miles away, it is nevertheless the first well-documented find of newly hatched ammonites and permits inspection of their embryonic shell features. These ammonite shells are about one millimeter in diameter and have a constriction at one end. They contain one relatively large buoyancy chamber in addition to the chamber that would have been occupied by the soft-bodied animal.

Ammonite embryonic shells are comparable in size to many calcareous plankton, and their large buoyancy chamber would have functionally equipped them to spend some part of their early life as plankton. This is a very common stage in the lives of many young snails and clams and for newly hatched squids and octopuses. As the ammonites continued to grow in size, they would have moved on to a more active mode of life in a variety of habitats. However, in their early life as plankton they would have succumbed to the general extinction of those surface-water organisms. The newly hatched and juvenile shells of ammonites preserved at the top of the Maastrichtian strata in Denmark may record this extinction. The contemporaneous nautilids that hatched at about ten times the size of the ammonites and actively swam in deeper water would not have been affected by that particular catastrophe.

These two potential explanations of the differential survival of ammonites and nautilids—specialist versus generalist and ammonite as plankton—may not provide science with a complete answer. Nevertheless, they may encourage further consideration of the complexities that determine survival—and extinction—in the biotic realm.

24

Dinosaurs' Lucky Break

MICHAEL J. BENTON

Throughout the history of life, there have been many major upheavals in which whole groups of animals were replaced by others. Perhaps the most famous was the replacement of the dinosaurs by the mammals 65 million years ago. Another major change occurred 150 million years earlier, when the dinosaurs took over the position of dominance that had been held for 80 million years by mammal-like reptiles.

What triggered these great upheavals? Scientific thinking about the various factors that might have been responsible for such replacements has undergone some significant changes of its own. At times, for example, scientists have suggested that mammals caused the extinction of the dinosaurs by eating their eggs or by competing for the same food resources. Now, however, most people are convinced that the mammals played only a minor role, if any (largely because both groups had lived side by side for millions of years), and that, instead, environmental change was primarily responsible. The mammals apparently sat around for 150 million years in the Mesozoic undergrowth until, with the extinction of the dinosaurs, they had their opportunity to radiate into newly available niches.

But what about the initial radiation of the dinosaurs? The transition from mammal-like reptiles to dinosaurs has been described by many authors and is frequently quoted as an example of major competitive replacement among vertebrates. (The use of the word *competition* in this context is convenient but potentially misleading. Individuals can compete, and one could even consider two species as competing, but at higher taxonomic levels, any large-scale replacements would probably be the result of numerous such events combined, together with other factors, in an essentially random way. The net result could more accurately be described as differential survival of one taxon compared with another that occupied a similar range of niches. For the purposes of this article, however, I will stick with the more familiar term of competition.)

According to nearly all published accounts, the replacement of mammal-like reptiles by the dinosaurs went as follows: During the Permian period and the early part of the Triassic (from 290 to 225 million years ago), the land was dominated by the mammal-like reptiles, which ranged in size and shape from a rat, or smaller, to a hippopotamus. At the end of the Permian, major upheavals occurred and many lineages became extinct. In the early Triassic, faunas of herbivorous mammal-like reptiles, almost certainly unbalanced by these events, were heavily dominated by huge numbers of a single genus, *Lystrosaurus*. New faunal elements also appeared at this time: the thecodontians, which included the ancestors of the dinosaurs. The carnivore niches throughout most of the Triassic were shared by advanced mammal-like reptiles, such as *Traversodon*, and thecodontians, some of which, such as *Rauisuchus*, became very large. The thecodontians are supposed to have gradually outcompeted the mammal-like reptiles and taken over the carnivore niches. Toward the end of the Triassic, there were still some herbivorous mammal-like reptiles, such as *Dinodontosaurus*, and some herbivorous thecondontians arose. Dinosaurs, according to most of these accounts, first appeared on the scene in the middle Triassic. Eventually, the mammal-like reptiles and thecodontians waned and the dinosaurs increased in importance, supposedly as a result of competition. The replacement was complete after about 40 million years.

Dinosaur superiority in this alleged competition has been variously attributed to features of their posture and locomotion or to their thermoregulation. Mammal-like reptiles and thecodontians are said to have had a sprawling or semierect posture, with their legs stuck out sideways and bent downward only at the knee and elbow. In dinosaurs, the legs were tucked right under the body, giving them an erect posture, as seen in most mammals. This enabled them to take longer strides and support greater weight.

In the past decade, many authors, including myself, have devoted hundreds of pages to speculations about the thermal physiology of the dinosaurs and why it was better than that of mammal-like reptiles. Impassioned arguments have been advanced for fully hot-blooded dinosaurs (with true endothermy as in mammals and birds), for cold-blooded dinosaurs (with true ectothermy as in living reptiles), or lukewarm-blooded dinosaurs (something in between: ectothermic, but with relatively constant internal temperatures because of their great bulk). These speculations on the locomotion and thermoregulation of dinosaurs are exciting, but I believe that the search for particular reasons for the competitive superiority of dinosaurs is fruitless. There is simply no evidence that any large-scale competition ever took place, and even if there were, animals are too complex for us to say that one or another adaptation can by itself explain a major worldwide ecological replacement.

How *does* one set about testing what happened 200 million years ago in order to identify the kinds of ecological forces that might have been operating? As with most broadly based studies of this kind, the route by which paleontologists reach their conclusions is convoluted and guided by accident as much as by inspiration.

I carried out my doctoral thesis work on a curious animal called *Hyperodapedon*, known from fossils found near the town of Elgin in northeast Scotland. *Hyperodapedon* lived some time before the end of the Triassic and had very close relatives, such as *Scaphonyx,* living at about the same time in India, Brazil, Argentina, Texas, Nova Scotia, and possibly Tanzania. These animals are rhynchosaurs—pig-sized herbivores with hooked snouts, slicing jaws with broad batteries of teeth, wide skulls with powerful jaw muscles, and hind limbs adapted for scratch digging.

I quickly found out that rhynchosaurs were *animalia non grata* in paleontological texts—neither mammal-like reptiles nor thecodontians nor dinosaurs, they did not fit the simple competitive models. But they could not be ignored, for when rhynchosaurs were present in fossil faunas, which was quite often, they were dominant. In fact, they made up approximately half of all reptile specimens found. Their relative abundance in the middle and late Triassic must have been comparable to that of antelope in certain African savanna faunas today. I was surprised, then, to learn that rhynchosaurs disappeared suddenly worldwide. I set out first of all to try to explain this extinction, but my search soon broadened in scope.

I wanted to have as accurate and localized an impression as possible of changes over time in the relative composition of late Permian and Triassic terrestrial reptile faunas. This meant that I had to collect information on the numbers of specimens of each species present in each of many separate faunas worldwide. To do this, I combined the few existing published accounts with visits to many museums—to count skulls and skeletons—and correspondence with curators in many more.

One factor facilitating my search for patterns in the transition from mammal-like reptiles and thecodontians to dinosaurs was the remarkable similarity worldwide among faunas at any particular time. During the Permian and Triassic, all land masses were a part of the supercontinent Pangea, and large land animals could migrate relatively freely. There was also apparently far less variation in temperature and climate from the equator to the poles than there is today.

The results of my study suggest a different sequence of events from those expected in the standard competitive model. Over the course of the Triassic, the carnivorous thecodontians did "take over" from the mammal-like reptiles, and this gradual change would seem to suggest that competition might have been involved. However, thecodontians never completely replaced the carnivorous mammal-like reptiles, which were radiating and evolving throughout this time. Indeed, among the mammal-like reptiles that persisted were many advanced forms (cynodonts), which gave rise to true mammals at the end of the Triassic. Both the carnivorous mammal-like reptiles and thecodontians died out toward the end of the Triassic and were replaced abruptly by dinosaurs. Thus, the fossil record of carnivores offers, at best, only weak support for an even partly competitive model.

Among the herbivores, mammal-like reptiles (particularly *Lystrosaurus* and other dicynodonts) were dominant in the late Permian and early Triassic. In the middle Triassic, the dicynodonts were partially replaced by other mammal-like reptiles (diademodontids) and by the rhynchosaurs. This situation continued until a few million years before the end of the Triassic, when all rhynchosaurs and major mammal-like reptile groups became extinct. A short time afterward (by geological standards), we find dinosaurs dominant worldwide. Again, the suddenness of the replacement seems to rule out competition as an important factor.

Dinosaurs such as *Staurikosaurus* had been present in the late Triassic rhynchosaur-diademodontid faunas, but only as rare elements (± 1 percent) and never large in size. Then, within one to two million years, the dinosaur radiation produced a range of different forms from modest bipedal carnivores (3 to 6 feet long) to large herbivores/omnivores such as *Massospondylus* (15 to 45 feet long). This changeover may be seen in Germany and the United States (Texas, New Mexico, Arizona), where successions of reptile faunas span the late Triassic and show "predinosaur" and dinosaur assemblages.

At this point a few words about the so-called middle Triassic dinosaurs— which are discussed in most texts and which are so important in the competitive model of replacement—are in order. Most of these "dinosaurs" were described earlier this century on the basis of single teeth and other fragments. More than half have since been shown to belong to the thecodontians and various common aquatic groups of the middle Triassic. Others are literally impossible to identify as anything in particular because they are so incomplete. The remaining dinosaurs from the middle Triassic or early part of the late Triassic have apparently been wrongly dated. They are all of late Triassic age, as shown by comparison of associated fossils with those from others parts of the world. Thus, I believe that there are no dinosaurs known before the late Triassic.

A close examination of the data, then, suggests that the dinosaurs came on the scene rapidly and that their initial pattern of radiation may have had much in common with that of the mammals 150 million years later. These opportunistic replacements, in which the new forms occupy niches that are already empty, are quite different from competitive replacements, in which the replacing forms actually cause the extinction of the replaced animals. And the difference is more than just an interesting aside to any particular changeover in the history of life: in a model of competitive replacement, one is stating that the new forms are somehow better than the old ones. In a model of opportunistic replacement, this need not be the case, and we do not have to assume that dinosaurs were superior to mammal-like reptiles and rhynchosaurs in any way at all.

But why *did* the mammal-like reptiles, rhynchosaurs, and thecodontians all die out toward the end of the Triassic? The extinction of major groups of terrestrial vertebrates—particularly the dinosaurs—has attracted a great deal of attention. The number of explanations for the extinction of the dinosaurs is getting out of hand, and not surprisingly perhaps, there is often a close tie between the ideas suggested and the discipline of the scientist making the suggestion. Thus, astronomers and geophysicists go for asteroids or comets; atmospheric scientists go for acid rain; ophthalmologists for cataract blindness; botanists for alkaloid poisoning; and dieticians for a reduction in fiber and natural oils leading to rampant dinosaurian constipation. In comparison with these, I have a somewhat less dramatic proposal to "explain" the extinction of the mammal-like reptiles, rhynchosaurs, and thecodontians: a change in the vegetation.

During the middle of the Triassic, southern parts of the world were dominated by the *Dicroidium* flora (seed ferns, horsetails, ferns, cycadophytes, ginkgos, and conifers). At the time of the reptile extinctions, this vegetation was largely replaced by conifers and bennettitaleans (large treelike cycads). The rhynchosaurs appear to have been highly specialized to the *Dicroidium* flora. Their jaws were designed for powerful and precise vertical bites, in which the lower jaw chopped into a deep groove in the upper jaw, like the blade of a penknife closing sharply into its handle. This remarkable jaw mechanism was an adaptation for cutting some of the tough vegetation found in the *Dicroidium* flora, and the loss of these food plants may have led to the animals' demise. (It is not clear whether some of the mammal-like reptiles and thecodontians were similarly specialized.) Coupled with this is a suggested climatic change: the first dinosaur-dominated faunas have been found in highly oxidized, reddish deposits, which are typical of a more arid climate and may indicate a general drying up. Perhaps the early dinosaurs were the only animals able to cope with the new vegetation and the hotter, drier climate.

If this opportunistic model for the rise of the dinosaurs proves valid, then, in combination with the later, equally opportunistic rise of the mammals, it raises an important question for paleontologists to tackle: have large-scale replacements *ever* been produced by competition? Several criteria might be looked for in the fossil record of major replacement events to determine whether they were opportunistic, competitive, or something else. Competition between one group and the group that replaced it would be suggested if, for example, the following patterns were found: as the replacing group increases in abundance over time, the other decreases; the rate of replacement is slow (for higher taxa, say, over a period of more than one million years); the two groups are found together and either can be dominant; and the replacement need not be associated with a climatic or floral change.

An opportunistic replacement, in contrast, might be suggested by the following patterns: the replacing group appears or radiates only after the extinction of the other; the rate of replacement is rapid (well below one million years); the two groups are not found together or, if they are, the new group will be unobtrusively present when the other is dominant; and the replacement is associated with environmental change.

Of course, a third kind of hypothesis to explain a replacement would be that it was neither opportunistic nor simply competitive but was caused by a mixture of different events—effectively random since we cannot define every factor involved.

Recently, a number of well-known, supposedly competitive replacements have been studied carefully, and they have turned out to be opportunistic. Stephen Jay Gould and C. Bradford Calloway of Harvard's Museum of Comparative Zoology looked at the replacement of brachiopods by bivalves (clams, for example) in the seas from the Paleozoic to the present day. Brachiopods were abundant, diverse, and widespread throughout most of the Paleozoic but declined afterward and now occupy a limited range of sea-floor habitats. Bivalves, which were already present in the early Paleozoic seas, increased in abundance and diversity and "took over" in the Mesozoic. Considered cursorily, this replacement appears gradual and, as such, a candidate for competition. However, when Gould and Calloway put together a simple graph of the numbers of brachiopods versus clams and other bivalves over time, they showed that the switchover was rather rapid and that it occurred at the boundary of the Permian and Triassic periods. At this time—the same time that many mammal-like reptiles were dying out on land—major extinctions swept through most marine groups. Whatever the cause of the wave of extinctions, the bivalves recovered from the disruption and radiated in the

Triassic, while the brachiopods continued but at reduced levels of diversity ever after. Far from being a long-term competitive takeover, this was opportunism.

Another recent subject of study has been the "Great American Interchange" of mammals that occurred when the Panamanian land bridge rejoined North and South America three million years ago and allowed free mingling of previously isolated faunas. The standard view of the interchange has been that the North American mammals made a greater impression on South American faunas than vice versa. This view concedes that a few South American forms (opossums, armadillos) made it to the United States and Canada, but the traffic was supposed to be primarily one way, from north to south. The clear implication is that North American forms were generally superior to South American ones.

Detailed studies by Larry G. Marshall of Chicago's Field Museum of Natural History and others (described by Stephen Jay Gould in the August 1982 issue of *Natural History*) have shown that there was no clear, one-sided competitive replacement of South American groups by North Americans. Many South American groups were little affected by the immigration, and many South American mammals—including tree sloths, monkeys, porcupines, anteaters, and various large rodents—successfully established themselves in the warmer parts of Central and North America.

The assumption that the succession of life forms through time is a manifestation of some principle of continuous improvement is pervasive and deep-rooted. As one group becomes outmoded, it is seen as inevitably replaced by a more dynamic, up-to-date assemblage. This is one view of human history, too. A story can be made up for any "competitive" replacement that we like. I believe, however, that the 150-million-year rule of the dinosaurs was not inevitable, that it could even be seen as an interlude between the two periods of evolution of the mammal-like reptiles and their descendants, the mammals. By any standard, the late Triassic mammal-like reptiles were advanced and diverse. True mammals appeared at about that time too—just after the origin of the dinosaurs. If the extinctions of most mammal-like reptiles (as well as rhynchosaurs and thecodontians) had not occurred 125 million years ago, there is every reason to suppose that the mammals could have radiated and come to occupy a broad range of niches during the end of the Triassic and the subsequent Jurassic period.

Recently, Dale Russell of the Canadian National Museum in Ottawa speculated about possible advances the dinosaurs might have made if they had not become extinct 65 million years ago. He proposed that certain relatively large-brained bipedal forms—the ostrich dinosaurs—might have become brainier and hominidlike. They could have been the dominant intelligent life form today. Another proposal, mind-boggling in some respects but less so in others, is that a mammal very like man might have arisen 150 million years ago if the dinosaurs had not had the opportunity to establish themselves after a large-scale extinction event.

PART 7

EVOLUTION AND SOCIAL ISSUES

The theory of evolution was born amidst a great outcry of controversy. Nor was the uproar strictly within scientific circles: on the contrary, scientists in general were quick to embrace the general notion that life has had an evolutionary history (though many had trouble with Darwin's proposed mechanism, natural selection). Most of the fuss stemmed from the perception that evolution was in conflict with the Judeo-Christian story of the origin of the world and its organic inhabitants. Darwin himself was aware that his materialist view of the history of life would trouble many, and his concern over this issue was probably the main cause of the great delay between coming up with the idea of evolution (late 1830s–early 1840s) and his publication of it in the mid-and late 1850s.

Prior to Darwin, the Genesis story of creation was not only standard cosmology for society in general, but also the basis for informed investigation of the earth, including its fossil record. The geologists who unraveled the order of strata in the upper part of the earth's crust were all creationists. After Darwin, most mainline Christian and Jewish sects ceded to science the role of explaining the origin and history of the cosmos, including life; no longer, for the most part, was it the obligation of religion to perform that role, nor did a particular story of origins seem required as the basis of moral teachings. But wherever the literal truth of the Torah or Bible remained a cornerstone of faith, science remained in direct conflict with religion.

Protestant Fundamentalism is the best-known, and, in the United States, most obvious creed emphasizing the literal truth and inerrancy of the Bible. The famous Scopes trial in 1925, usually remembered as the triumph of modern sophistication over rural backwardness, was actually a victory for creationists: evolution all but disappeared from most high school biology texts, not to reappear until *Sputnik* revealed the sorry state of science education in the United States. But just as evolution, and science in general, was making an important comeback in the curriculum of public schools, "special" or "scientific" creationism was being developed by a few representatives of a creationist tradition that has never entirely left the American scene. No longer trying to ban evolution from the schools, but looking, instead, for "equal time" for a "scientific" version of creationism, the current versions of creationism are themselves far more sophisticated than their earlier counterparts. And though, in the mid-1980s, creationism is once more on the wane, it has merely reverted back to the local level, with activists still very much determined to see the old biblically based cosmology injected into the science curriculum. The main danger of this effort is that our children will become more confused than ever about what science *is*—not an appetizing prospect as the world becomes ever more industrialized and dependent upon science and technology. We simply cannot afford to have a scientifically illiterate electorate.

Though the Scopes trial produced little effect in the pages of *Natural History* (despite the fact that the American Museum's president, Henry Fairfield Osborn, was on the list of prospective scientific witnesses—none of whom were ever called), later developments were reflected in the magazine. Norman D. Newell, now a Curator Emeritus at the American Museum, was for many years among the very few in the scientific community to voice his concern over the creationist threat. Interweaving a cogent account of the development of historical science, most especially evolution, with his account of creationism circa the early 1970s, Newell outlines the nature of the "special creationism" attack. Back then, Newell claims, creationists were less concerned with attacking evolution than establishing their own views on an equal footing.

Not so in the early 1980s, as Laurie Godfrey tells us in her portrayal of creationist tactics in this, the most recent (but, alas, assuredly not the last) creationist uprising. Godfrey documents how creationists distort the writings of scientists, and seek to establish the weakness of the evolutionary position by pointing to instances where scientists openly disagree with one another. But, as Godfrey reminds us, scientists are *supposed* to disagree—otherwise there can be no growth of scientific knowledge and understanding. In modern "creation science," evolutionary biologists stand guilty—of doing science.

In the final essay, noted evolutionary biologist John Maynard Smith steps back and examines the relationship between myth and science. Admitting that, as a youth, he was in favor of rejecting all myths for the hard-nosed realities afforded by science, Maynard Smith now sees a role for all manner of stories that we may tell ourselves to help us define the human place in this world. And, as his career so ably demonstrates, he has also retained his conviction of the importance of science. In this essay, Maynard Smith is concerned to distinguish between myth and science, and to argue that myths should not be based on science, nor science forced to fall into line with myth. Each deserves its own place under the sun. For my own part, I would agree; realizing that life has had a history—has *evolved*—has helped us learn much about the organic world. To build a moral or ethical philosophy upon this edifice, as some have tried to do from time to time, seems inherently destined to failure. It is the scientist's task to describe the natural world as accurately as possible. But it is the job of us all to come to grips with the nature, essence, and "meaning" of human existence.

25

Evolution Under Attack

NORMAN D. NEWELL

An old controversy concerning the origins of life and biological change is being quietly rekindled. The special creationists, who believe in the literal truth of the biblical account of creation despite the contrary evidence of astronomy, geology, paleontology, and genetics, are once again opposing the evolutionists. The issue is how the history of life should be taught in public school science classes.

Organic evolution was publicly challenged in this country in the Scopes trial of 1925 in Dayton, Tennessee. Although Scopes lost this test case, the fundamentalist position, as expounded by William Jennings Bryan, was made to appear so untenable that it was presumed by the scientific community and the public at large that special creation would never again be an issue in public education. That conclusion was premature.

In 1971, the Board of Education in Columbus, Ohio, passed a resolution encouraging teachers to present special creation along with evolution in discussions of the origin of life or the universe. After nine years of petitioning, the California State Board of Education ruled in 1972 that henceforth science textbooks must teach evolution as a theory rather than a fact, and that all statements on Darwinism must be carefully qualified. This was a compromise. What the fundamentalist lobbyists had sought was equal time in science teaching for the theory of special creation. More recently, the board adopted a resolution to require equal recognition of creationism in social science textbooks wherever evolution is discussed. And efforts are still being made in California to compel publishers to join creation and evolution in biology texts.

In 1972, William F. Willoughby, religion editor of the *Washington Star-News,* brought suit as a private citizen in U.S. District Court against the National Science Foundation and its Biological Sciences Curriculum Study to force withdrawal of NSF-sponsored biology textbooks on the grounds that they present the

Darwinian theory of the origin of man as fact, not theory, and omit mention of special creation. The NSF has estimated that about 45 percent of today's high school students study from BSCS texts, which were introduced in 1960. In 1973 the suit was dismissed. A bill requiring equal time for special creation in science classes is presently pending in the Georgia State Legislature. Similar legislation was passed in Tennessee in 1973 but was defeated in Michigan, Colorado, and Washington.

The National Association of Biology Teachers is deeply concerned that the academic freedom of public school teachers is being threatened and the basic principles of science are being attacked. The association is presently combating the creationists in the Tennessee courts and expects to carry its case as far as the U.S. Supreme Court if necessary. The biologists argue that the new Tennessee statute contravenes the free speech clause of the first amendment and the due process clause of the fourteenth.

Professional biologists in various parts of the United States have protested the lobbying tactics of the creationists, and several of the nation's academic and scientific organizations have passed resolutions condemning the creationists' proposals on the grounds that special creation is theology, not science, and should not be represented to public school children as a reasonable scientific alternative to the theory of organic evolution. Included among these organizations are the American Association for the Advancement of Science, the American Chemical Society, the University of California Academic Senate, and the National Academy of Science. The last-named, the most distinguished scientific body in the United States, was established by Abraham Lincoln to advise the government on scientific matters.

Who are the special creationists? Primarily, they are members of small, independent, predominantly fundamentalist groups. Many belong to evangelical churches whose total United States membership is estimated at about twenty million. Fundamentalists are not a monolithic assemblage; their viewpoints range from conservative to progressive, and they often find no conflict between science and religion. Perhaps the most vocal of the antievolutionist groups is the Creation Research Society of San Diego, which has developed its own textbook, *Biology: A Search for Order in Complexity,* and whose members were active in the 1972 California ruling. Founded in 1963, this society, which currently numbers about 300 full members and 1,200 sustaining and student members, has two requirements for voting membership. Applicants must hold an earned advanced degree in a recognized area of science and, like the nonvoting members, who need not possess an M.A., M.S., or Ph.D., they must subscribe to the society's credo, which states in part that the account of origins in Genesis is a factual presentation of simple historical truths.

All basic types of living things, including man, were made by direct creative acts of God during Creation Week. . . . Whatever biological changes have occurred since Creation have accomplished only changes within the original created kinds.

The Noachian Deluge was an historical event, worldwide in its extent and effect.

These tenets are a throwback to the frankly scriptural explanation of living forms that prevailed in the first half of the nineteenth century. They echo the then held belief that all life came into being separately and spontaneously, that species are discrete and immutable, that nature is static, and that consequently all forms that would ever exist were already in existence. At the time it was also widely accepted that the universe and man were created in the year 4004 B.C., a date derived by Archbishop James Ussher, a seventeenth-century Irish prelate and scholar, from his study of biblical genealogies.

A very different view of nature is held by the evolutionists. They think in terms of a world of enormous antiquity and gradual, continuous change in which existing plants and animals, including man, have slowly evolved from previously living forms. They also believe that the process of evolution is still going on.

Evolutionary thinking began to appear many years before Charles Darwin formulated his theory. Darwin's grandfather, physician-naturalist Erasmus Darwin, wrote in 1794, "Would it be too bold to imagine that in the great length of time since the earth began to exist, perhaps millions of ages before the history of mankind . . . all warm-blooded animals have arisen from one living filament?" And Jean Baptiste Lamarck, French naturalist and professor of zoology, said in the early nineteenth century, "Citizens, go from the simplest to the most complex and you will have the true thread that connects all the productions of nature; you will have an accurate idea of her progression; you will be convinced that the simplest living things have given rise to all others."

What Charles Darwin did in the *Origin of Species* was assemble and interpret the evidence for evolution and discover the principle of natural selection by which it operates. Through an historic coincidence, Alfred Russel Wallace, another English naturalist, had simultaneously and independently deduced the same principle. This is how Darwin described that part of his theory in the introduction to *Origin*, published in 1859:

As many more individuals of each species are born than can possibly survive [the doctrine of Malthus]; and as, consequently, there is frequently recurring struggle for existence, it follows that any being, if it vary however slightly in any manner profitable to itself, under the complex and sometimes varying conditions of life, will have a better chance of surviving, and thus be *naturally selected*. From the strong principle of inheritance any selected variety will tend to propagate its new and modified form.

Most biologists now recognize natural selection as the directive force in evolution. No modern evolutionist believes that evolution is the result of a long series of random accidents. Darwin, however, did not know the cause of the variations he had noted within species, nor how they were passed on to succeeding generations. That gap has been closed by the subsequent discovery of the mechanisms of inheritance and knowledge of genes, mutations, and the genetic recombination that always results from sexual reproduction.

Since evolution by natural selection was incompatible with the basic beliefs of many religious creeds, especially with the presumed place of mankind in the scheme of things, the *Origin of Species* caused an uproar throughout the Western world. It precipitated widespread polarization among the general public in Europe and America, separating those who insisted on the literal interpretation of the book of Genesis as a historical account from those who regarded Genesis as a poetic legend. The controversy was seen as pitting science against religion. It was dramatized in the 1860 debate between Bishop Samuel Wilberforce, upholding Genesis, and Thomas H. Huxley, defending Darwin, during which the bishop asked his opponent whether it was through his grandfather or his grandmother that he claimed descent from an ape.

Although the theories of evolution and special creation are essentially the same today as they were in Darwin's time, the latter-day creationists have not mounted a frontal attack on evolution and, for constitutional reasons, they carefully avoid mention of Genesis in their proposals for the public schools. Instead, they assert that the creation theory is as valid as evolution and therefore deserves equal attention in schoolrooms, and that the dispute this time around pits science against itself. Full members of the Creation Research Society hold, after all, master's and doctoral degrees in science.

Nevertheless, as the time approached for the California Board of Education to make its 1972 decision, nineteen resident Nobel laureates sent a letter of protest in which they said, "No alternative to the evolutionary theory gives an equally satisfactory explanation of the biological facts." At a juncture when science and technology have split the atom, cracked the genetic code, and put men on the moon, the current revival of pre-Darwinian theory has an eerie, dreamlike quality.

The authoritarian and apparently impregnable cosmology of the Christian

Middle Ages was overthrown in the sixteenth and seventeenth centuries by Copernicus, Bruno, Kepler, and Galileo, and the descriptive mechanics of celestial bodies was established by Newton. These men did not, however, conceive of a long prior history of gradual change in the universe. Time is crucial to the process of evolution. The long, slow change outlined by the evolutionists required a far greater antiquity for the earth than the orthodox 6,000 years. Based on the minimum amount of time needed to lay down and then denude ancient layers of sediment in southern England, Darwin himself estimated the age of the earth at millions of years. In this century, that figure has undergone a massive upward revision. The universe is now generally believed to be about ten billion years old, and the earth about five billion. Life is believed to have originated on our planet some three or four billion years ago. That is ample time for the drama of organic evolution.

The work of early geologists and the discovery and interpretation of trends in the form and structure of fossils played an important role in the development of evolutionary thinking.

The basic principles of earth history are illustrated by the work of many brilliant scientists, spanning the period from the seventeenth century to the early 1900s. Among them were Nicolaus Steno, James Hutton, William Smith, Charles Lyell, and Darwin. They opened the way for historical interpretations of the origins of earth and its life and the evolution of stars and planets.

Steno's contribution appeared in 1669. He developed his theory of geologic development through his studies of crystal growth and larger features such as layered sedimentary and volcanic rocks. He demonstrated that these phenomena are the results of sequential changes, that they have histories. Sedimentary strata now cover the sea floor and much of the land area—in fact more than 90 percent of the earth's surface—reaching a thickness in places of several miles. They contain the fossil remains of bygone plants and animals and other records of a long history of past conditions and events. Their present average rates of accumulation are well known. These rates, of course, vary greatly from place to place, but they are consistent with the conclusion that many hundreds of millions of years were required to erode the lands and deposit the resultant material to form the sedimentary record. This record is made accessible by deep borings and deep erosion, especially in mountains where miles-thick sequences of sedimentary layers are often exposed to direct examination.

The second milestone in the formulation of historical science was the publication, in 1796, of *The Theory of the Earth,* an epochal work by the Scottish naturalist James Hutton. Hutton independently rediscovered Steno's method of analyzing the evidence of sequential steps in past events. He adopted the Newtonian view that God has always acted through laws and that the majestic works of nature glorify Him and attest to His existence. He rejected the possibility of repeated divine control of events as undemonstrable, speculative, and unnecessary. Instead, he showed that known physical causes—gravity, erosion, sedimentation, heat and cold, volcanic eruptions, and earthquakes—are adequate to explain geologic features. His reasoning led him to conclude that the known rates of change, although admittedly variable, would require much more time to build the earth than the 6,000 years allowed by theologians. Geologists have come to think of this orderly application of natural laws in geologic evolution as *uniformitarianism,* or simply as uniformity of process—the present is the key to the past. A modern corollary is that the past can teach us much about the present and the future.

Hutton's principle was adopted and broadly illustrated by Charles Lyell, a compatriot and contemporary of Darwin, in his three-volume *Principles of Geology,* which became a standard work on stratigraphical and paleontological geology. While lying seasick in his hammock on the outward-bound Beagle, Darwin read the first volume, published in 1830, in which Lyell argued that the earth's contours were not shaped by the Flood but by the action of rain, wind, earthquake, volcano, and other natural forces. "No causes whatever have from the earliest time to which we can look back, to the present, ever acted but those now acting and they have never acted with different degrees of energy from which they now exert," Lyell wrote.

Fundamental clues concerning past life and earth history were also provided by William Smith, an English surveyor, engineer, and practical field geologist, who discovered that changes in kinds of fossils could be used to determine the sequence and relative ages of the deposits in which they were found. In 1815 he published the first geologic map of England, which showed that each individual stratum contained "organized fossils peculiar to itself." The strata, he found, were arranged in a particular and fixed sequence over large areas. This discovery, subsequently confirmed over the world, makes it possible to classify strata by geologic age and to date past events in chronological order. These facts of the fossil record are satisfactorily explained by evolution, not by the creation idea.

Geologists before Darwin's day thus learned to identify the geologic ages of strata by their distinctive assemblages of fossil remains, especially those of extinct but abundant marine organisms. The outline of a worldwide system of classification of geologic strata and their ages based on the vertical sequence was completed early in the nineteenth century and is now routinely employed for geologic mapping, the search for mineral deposits, and for the correlation of past events in the physical and biological evolution of the earth. All of this development depended on worldwide changes in life.

At about the time Smith was making his discoveries, Georges Cuvier, a French naturalist, was busy classifying fossil remains unearthed in excavations in and around Paris. Based on what was known of comparative anatomy, he organized ancient shells and bones into a chronological succession of genera and species, among which he counted about ninety that had disappeared from the earth. What had caused their extinction? Unable to reconcile his finding of successively vanished worlds with the dictates of fundamentalist principles and unwilling to abandon his faith in special creation, Cuvier embraced the doctrine of catastrophism, a popular nineteenth-century theory, which deemed that it was not the Noachian Deluge alone but a series of cataclysmic events that reshaped the earth and destroyed former species. According to Cuvier's version of catastrophism, extinct species were replaced by migrations from other regions. Other contemporary catastrophists, impressed by consecutive changes in whole faunas and floras, believed that each catastrophe was followed by new acts of creation. Many of today's creationists are even more conservative than their nineteenth–century forerunners in their insistence on a single burst of creation and a single catastrophe—the Flood.

Hundreds of alternating assemblages of land and sea organisms, of which all but the youngest are extinct, have been traced over the world. They tell not simply of one flood, but of many advances and retreats of the sea over portions of each of the continents. Buried remains of still-living species of animals and plants, although quite numerous in the earth, are found only shallowly buried within a few feet of the present surface in still unconsolidated sediments. They form a very small and inconsequential part of the fossil record. This agrees with the conclusion of paleontologists that the modern biota does not extend back into the deeper strata believed to be more than 25 million years old. If these species had been created together with all other forms of life, as the creationists say, they would occur throughout all abundantly fossiliferous strata—representing 600 million years of earth history—from bottom to top. They do not. The thousands of fossils species that extend through vertical miles of strata below the records of still-living species belong to wholly extinct species of organisms that show successive vertical changes through the rocks. They include many forms that clearly are to be classified in the same major groups with modern organisms of which they are presumed to be the ancestors.

For several decades, geochemists have been dating events of earth history by means of the constant rates of decay of radioactive isotopes. This is the radiometric method. Carbon 14 is effective in dating objects up to 50,000 years old. Using radioactive isotopes of uranium, thorium, potassium, rubidium, and other elements the time scale can be extended back far beyond the 600 million years of the abundantly fossiliferous part of the record.

As radiometric methods have been refined and combined with other methods of relative dating provided by rocks and fossils, overall precision and reliability have steadily improved. An accuracy of plus or minus 10 percent of the age, in years, is routine for good samples. The radiocarbon method of dating has recently been greatly improved by cross-checking radiocarbon analyses with overlapping suites of annual growth rings of trees.

Among the phenomena that impressed Darwin, as it had other naturalists engaged before him in classification of organisms, was the variation in form within single populations and the underlying similarity of structure among quite different species. To the creationists, variation, as explained in the biology textbook of the Creation Research Society, "is simply an expression of the Creator's desire to show as much beauty of flower, variety of song in birds, or interesting types of behavior in animals as possible," presumably for the delectation of human beings. Similarities are attributed to the fact that "the Creator chose to use a common pattern" in his separate acts of creation. Organic evolutionists, on the other hand, attribute variations to genetic recombinations and mutations. Similarities in form, structure, and DNA proteins are believed to indicate a common origin and the descent of one species from another.

Much chemical and genetic evidence exists of the essential unity of life on earth. For example, one or both kinds of nucleic acid, DNA and RNA, by which heredity is encoded, stored, and retrieved, carries the genetic message in every organism from viruses and bacteria to man. These complex nucleic acid molecules are paired spiral chains of nucleotides wound around a common axis to form a double helix. The nucleotides vary in sequence within the chains. The number of nucleotide pairs in a single gene, or hereditary factor, may range into the thousands, giving the genes the potential for a variability so immense that it probably exceeds the realized diversity of all the organisms that have ever lived on earth.

Only four kinds of nucleotides—or genetic "letters"—are employed to spell out the genetic code—or "words"—of the chromosomes. Evolution has been guided by the formulation of new words, created by sexual reproduction or recombination and mutation, that pass the test of natural selection.

In rejecting the concept of evolution *in toto*, the special creationists, who still maintain that the earth is only 6,000 years old, level many criticisms against it. Chief among them are these: enough time does not exist for major evolutionary changes to have taken place; gaps between major groups exist in the fossil record;

there is a tendency for whole groups of species to appear suddenly, contrary to what one would expect in evolutionary development; the orderliness of nature could not come about by natural means; and, finally, organic evolution is not a fact but only a theory that has never been proved.

Estimates of the earth's age and the methods of geologic dating have already been explained. Lacunae and aberrations in the fossil record are answered by the explanation that the earliest population of any new group consists of very few individuals, and the chance of their being preserved is therefore negligible. The order of nature is far from perfect. What order exists, conforms to the action of known natural laws.

As to the last criticism, there is no argument. Evolutionists willingly concede that their explanation of evolution is a "theory." It is, however, an organized explanation, a conceptual scheme that satisfies the facts as presently known. The evidence on which evolutionary theory is based is indeed factual, overwhelming in quantity and variety, and abundantly proved. If organic evolution is defined as profound changes in organisms throughout the history of life, then evolution may be accepted as a fact. But like everything else in science, the theory of organic evolution is open to revision should new evidence appear.

The evolutionists, for their part, dismiss special creation not only because it flies in the face of centuries of accumulated knowledge but because it violates the spirit of science by substituting supernatural explanations for empirical evidence. The doctrine of special creation is theology, not science. As the National Academy of Science put it in protesting the lobbying tactics of the special creationists:

> The result of including creationism in otherwise nonreligious textbooks would be to impair the proper segregation of the teaching and understanding of science and religion. The foundations of science must exclude appeal to supernatural causes not susceptible to validation by objective criteria. Science and religion being mutually exclusive realms of human thought, their presentation in the same context is likely to lead to misunderstanding of both scientific theory and religious belief.

To those who think that the battle over evolution was settled in the Wilberforce-Huxley debate in 1860, or in the Scopes trial in 1925, one can only repeat the words of philosopher-poet George Santayana: "Those who cannot remember the past are condemned to repeat it."

26

The Flood of Antievolutionism

LAURIE R. GODFREY

In 1963, American historian Richard Hofstadter wrote that "today the evolutionary controversy seems as remote as the Homeric era." The Biological Sciences Curriculum Study Project, supported in part by federal funds, was preparing secondary school texts that openly presented evolution as the foundation of biology. And George McCready Price, an outspoken leader of the protest against evolution in the days of the Scopes "monkey trial" and author of numerous antievolutionary tomes, including *The Phantom of Organic Evolution* (1924), *A History of Some Scientific Blunders* (1930), *The Modern Flood Theory of Geology* (1935), and *Genesis Vindicated* (1941), died at the age of 92. But 1963 was also the year that the Creation Research Society—and with it, organized "scientific creationism"—was born.

The Creation Research Society was founded by a group of ten men led by Walter E. Lammerts and William J. Tinkle. Many of these men were disaffected members of the American Scientific Affiliation, a theistic organization founded in 1941 and devoted to the reconciliation of science and evangelical Christianity. The increasing domination of the organization by evolutionists disturbed those who wanted it to oppose evolutionism. The "team of ten" vowed to work, through what they regarded as scientific endeavors, for a revival of belief in special creation as described in the King James version of the Bible. While they held populist William Jennings Bryan, the Scopes prosecutor, in high esteem, the new activists were creationists of a different ilk.

Bryan had mocked his scientific opponents: "You believe in the age of rocks; I believe in the Rock of Ages." He had preached to the masses, "I would rather begin with God and reason down then begin with a piece of dirt and reason up." But the new creationists profess no disdain for science. College-educated fundamentalist Christians with a strong commitment to both science (particularly in the form of technology and engineering) and to a literal interpretation of the

Bible, they have set out to convince the public that "true science" supports the creation model of world and life origins. Denying that they are trying to bring religion into the public schools, they are seeking to have their model taught as science.

By the end of 1980, seventeen years after Hofstadter had pronounced the evolution controversy dead, "two model" scientific education bills—which would require public schools to present creation as a scientific model alongside evolution—had been introduced and debated in the state legislatures of Florida, Georgia, Illinois, Iowa, Kentucky, Louisiana, Minnesota, New York, South Carolina, Tennessee, and Washington and were being introduced elsewhere. In addition, various local school boards around the country had passed resolutions that made concessions to creationist pressure. The membership of the Creation Research Society, based in Ann Arbor, had grown to 2500. Sister organizations such as the Bible Science Association (Minneapolis), the Creation Social Science and Humanities Society (Wichita), the Institute for Creation Research and the Creation Science Research Center (San Diego) had been formed to defend scientific creationism and promote the teaching of creation as equal with evolution.

Led by virtually the same nucleus of antievolutionists, these organizations have become efficient factories of purportedly scientific antievolutionary propaganda. Conventions, as well as debates, textbooks, and films, are the means to the political end of building a constituency. The Institute for Creation Research (ICR) now offers college and graduate-level programs as well as summer institutes (offering optional college credit) on creationism; distributes antievolutionary books, pamphlets, and cassettes; and sponsors creation/evolution debates and nationally distributed weekly radio broadcasts. And the ICR also funds research: to find evidence of Noah's ark and a global flood; evidence of coexisting humans, trilobites, and dinosaurs; and proof of a recent creation of the universe and the planet Earth (the earth is assumed to be roughly 10,000 years old). The Creation Research Society developed the first "creation science" biology textbook meant for use in public secondary schools, and since 1964 the society has published a quarterly journal dealing with evidence that supports a literal interpretation of the Bible.

The scientific creationists make no attempt to hide the proselytizing role of their various research organizations. Emphasis Creation 1980 was a Midwestern convention of scientific creationists sponsored jointly by the ICR and the Bible Science Association. The Director of the ICR, Henry Morris, gave explicit instructions, which appeared in the newsletter of the ICR's Midwest Center:

> The purpose of such a convention should not be to provide a forum where various creationists get together to present papers arguing for their own particular

interpretations on details of science or Scripture. Rather, it should seek to reach as large and general an audience as possible with carefully chosen papers (and other activities) by qualified speakers who will make the greatest impact for the creationist cause in general.

The newsletter went on to list acceptable and unacceptable topics. The former included refutations of evolutionism; legal, political, and educational aspects of teaching creation in schools; scientific evidence for a recent creation of the earth and universe; and "flood geology," which attributes a wide range of fossil-bearing geologic deposits to a single catastrophic global event, the Noachian deluge. Unacceptable topics included plate tectonics and continental drift (listed among other as areas of questionable or peripheral significance to creationism) and all "highly technical and specialized treatments of individual problems."

Field or laboratory research represents a very minor charge of scientific creationists. Most efforts are directed toward rewriting the discoveries and inter-pretations of evolutionists. In this endeavor, numerous evolutionists are portrayed as scientists who have all the evidence to disprove evolution (and support creation) at their fingertips, but who are either too stubborn or too deeply indoctrinated in evolutionary dogma to appreciate it. Arguments of anthropologists, biologists, chemists, geologists, astronomers, physicists, and engineers are reinterpreted or taken out of context. In this way, as I will show below, creationists manage, among other things, to convert arguments about the pattern and process of evolutionary change into arguments about the existence of change.

The primary tactic of the scientific creationists is to find controversy, disa-greement, and weakness in evolutionary theory—by no means a difficult task. Having demonstrated problems with various aspects of evolutionary theory (some fabricated, some real), the creationists then conclude that we must accept the Judeo-Christian biblical account of creation as the only possible, logical alternative. Thus scientific creationism proceeds by constructing an artificial dichotomy be-tween two models—evolution and creation—both incorrectly respresented as monolithic.

Actually, various evolutionary explanations are possible, and numerous models, both Darwinian and non–Darwinian, have been posed. They have in common the notion that the earth's life forms are related by common ancestry, whether or not they have since achieved reproductive isolation. Evolutionists agree that the evidence supports this premise of genetic continuity although, as scientists, they do not rule out the logical possibility that life could have arise independently on more than one occasion on the earth or in the universe.

Creationism, on the other hand, is based on the idea that reproductive isolation usually signals the absence of common ancestry. Given genetic discontinuity, numerous creation-based explanations are nevertheless possible: witness the global diversity of creation myths. Ignoring this diversity, however, scientific creationists begin with one specific and detailed explanation of the universe and require its acceptance on faith as a prerequisite of membership in their various research organizations. The Statement of Belief of the Creation Research Society begins: "The Bible is the written Word of God, and because we believe it to be inspired throughout, all of its assertions are historically and scientifically true in all of the original autographs." The scientific creationists do not pose and test alternative creation models. Doing science is not the business of scientific creationists; destroying the public credibility of evolution is their real goal. "New evidence," the press is told, reveals "major weaknesses" in evolution. Oddly, the creationist tactic of discovering that evolutionary biologists are guilty of doing science—posing, testing, and debating alternative explanations.

One scientific debate in particular, that between the neocatastrophists (or punctuationalists) and the phyletic gradualists, has fueled the fires of scientific creationism. In 1972 Niles Eldredge of the American Museum of Natural History and Harvard paleontologist Stephen Jay Gould launched their new theory of evolution by "punctuated equilibria." Evolution, they claimed, proceeds by rapid fits and starts, punctuating long periods of relative stasis. Drawing from the work of other paleontologists and neontologists, Eldredge, Gould, and later, Steven Stanley of Johns Hopkins developed the implications of a punctuational model of evolutionary change. In so doing, they challenged the assumption that most evolutionary change occurs as a byproduct of slow, ceaseless natural selection acting on variation in well-established populations.

While they have not abandoned the concept of natural selection as an important evolutionary process, the punctuationalists have reinterpreted its rule. Central to their argument is the view that most evolutionary change occurs in association with speciation, that is, the formation of independent species by the splitting of lineages into reproductively isolated populations. They argue that speciation may be achieved rapidly in small, geographically isolated populations and that, in such populations, chance (as well as natural selection) can exert much greater influence on genetic change in form than is otherwise possible. They further argue that rapid or dramatic evolutionary changes cannot occur in the *absence* of speciation. The mechanisms and importance of speciation lie at the heart of the debate between the punctuationalists and their opponents. Unlike the phyletic gradualists, the punctuationalists conclude that in macroevolution (evolution viewed in the long range and on a large scale), an episodic pattern of change is the expectation, rather than an exception to the rule.

A second important issue in evolution that has attracted the attention of creationists is the question of the relative importance of chance as a factor in evolutionary change. Using computer simulations, David Raup and his colleagues at Chicago's Field Museum of Natural History have argued that chance is very important in macroevolution as well as macroevolution. Raup believes that many genetic changes that do not greatly affect "fitness" may survive or fail to survive owing to chance. Such evolution by chance is called neutral, or non-Darwinian, evolution. The debate in evolutionary biology is over its relative importance, not its existence.

It is hard to imagine creationists drawn to the arguments of Eldredge, Gould, and Raup, since they are antithetical to creationist tenets. First, the question of the genetics of speciation, which is central to the theory of the punctuationalist school, is foreign to creationism. "Speciation" is rarely part of creationist vocabulary; "special creation" is used instead. Creationists claim that each life form was created as a separate "kind" (a common-sensical, undefined concept) and that nature permits variation only within such created kinds. Thus they must ignore a large body of biological data on the mechanisms of speciation and examples of its occurrence observed both in the laboratory and the field. Scientific creationists deny the existence of the very process that punctuationalists argue is critical to evolutionary change.

Nothing about punctuationalism supports the creationist viewpoint. Punctuationalists simply maintain that while much evolutionary change is very slow or static, very rapid "jumps" can occur naturally and these are the important stuff of macroevolutionary change. Genetically, such jumps are as comprehensible as slow phyletic changes. Indeed, whether they are perceived as jumps at all depends upon one's expectations concerning the scale and pace of evolutionary change. As Gould has written (*Natural History*, August 1979):

> New species usually arise, not by the slow and steady transformation of entire ancestral populations, but by the splitting off of small populations from an unaltered ancestral stock. The frequency and speed of such speciation is among the hottest topics in evolutionary theory today, but I think that most of my colleagues would advocate ranges of hundreds or thousands of years for the origin of most species by splitting. This may seem like a long time in the framework of our lives, but it is a geologic instant, usually represented in the fossil record by a single bedding plane, not a long stratigraphic sequence.

Second, "chance" is also foreign to creationism. One Florida-based organization puts out a flier that reflects the widespread creationist notion that nothing (or nearly nothing) ever happens by chance: "Evolution demands what has not, and cannot happen, even with careful planning—much less by total accident!" It is, of course, a misstatement of evolution to claim that this body of theory argues that change comes about "by total accident," for selection is not a random process. Yet, the non-Darwinian school ascribes to chance a much more central role than is admitted by other evolutionary biologists. Ironically, in their effort to show disagreement among evolutionists, the creationists are citing the work of paleontologists whose arguments are, in many ways, the most antithetical to creationism.

One reason creationists are able to exploit the current debates among evolutionists is that certain key phrases have entirely different meanings for paleontologists and for creationists (or their constituency). When such phrases are lifted from the work of evolutionists and inserted into creationist literature, they acquire new meaning simply because of differences in assumed knowledge. For example, the "neocatastrophism" of paleontology (widely quoted in support of creationist catastrophism) has nothing to do with either creation or a great flood. But creationists automatically associate the term "catastrophism" with the concept of the Noachian deluge.

Creationist Gary Parker wrote an essay on neocatastrophism that was circulated in the October 1980 issue of *Acts and Facts,* the free monthly newsletter of the ICR. Reading his article one cannot avoid the conclusion that Raup and Gould consider the creation model tenable, if not actually preferable to evolutionism. Here is a passage from Parker's essay:

> "Well, we are now about 120 years after Darwin," writes David Raup of Chicago's famous Field Museum, "and the knowledge of the fossil record has been greatly expanded." [Parker cites a 1979 article by Raup.] Did this wealth of new data produce the "missing links" the Darwinists hoped to find? "Ironically," says Raup, "we have even fewer examples of evolutionary transition than we had in Darwin's time. By this I mean that some of the classic cases of Darwinian change in the fossil record, such as the evolution of the horse in North America, have had to be discarded or modified as a result of more detailed information." Rather than forging links in the hypothetical evolutionary chain, the wealth of fossil data has served to sharpen the boundaries between the created kinds. As Gould says, our ability to classify both living and fossil species distinctly and using the same criteria "fit splendidly with creationist tenets." "But how," he asks, "could a division of the organic world into discrete entities be justified by an evolutionary theory that proclaimed ceaseless change as the fundamental fact of nature?" [Parker cites a 1979 *Natural History* article by Gould.] "we still have a record which *does* show change," says Raup, "but one that can hardly be looked upon as the most reasonable consequence of natural selection." The change we see is simply variation within the created kinds, plus extinction.

The arguments Parker presents outside, as well as inside, quotation marks seem to be those of Raup and Gould. Given these selected tidbits, there is no way to interpret the statements of Raup and Gould except within the framework of the creation model. The reader is not told what Raup and Gould are arguing but is left instead to surmise, incorrectly, that evolution itself is under attack. Furthermore, Parker has chosen to cite titles that seem to support such an interpretation. Raup's article is called "Conflicts between Darwin and Paleontology." Gould's is entitled "A Quahog Is a Quahog."

Those familiar with Raup's research will not be surprised to find that his article is actually a treatise concerning problems with Darwinian gradualism. Raup first deals with the complex, uneven record of evolutionary change. His point, quoted more fully, is that "some of the classic cases of Darwinian change in the fossil record, such as the evolution of the horse in North America, have had to be discarded or modified as a result of more detailed information—what appeared to be a nice simple progression when relatively few data were available now appears to be much more complex and much less gradualistic." Raup goes on to discuss the potential of chance processes to bring about apparently patterned evolutionary change—in particular, the extinction of lineages.

Gould's article is also about problems with Darwinian gradualism. It takes to task those biologists and anthropologists who argue that species boundaries are artifacts of the human capacity to classify and construct artificial divisions. Gould argues, as Ernst Mayr did years before, that species are real biological entities, but he does not suggest that they are genealogically unrelated to one another or that they cannot give rise to new species.

Despite the attempts of scientific creationists to play up the signs of controversy among evolutionists, there is actually widespread agreement in scientific circles that the evidence overwhelmingly supports evolutionism. Confirmation has sometimes taken unexpected forms, as in the high correlation between the degree of biochemical difference between pairs of species and the amount of paleontological time since their apparent separation.

There is agreement that the pattern of origin of taxa in the paleontological record strongly supports genetic continuity and, therefore, evolution. The punctuationalists' concept of evolutionary stasis has been misused by creationists to argue against such a pattern, but evolutionary stasis contradicts only strict gradualism, not evolution. The fact is, the genus *Homo* does not occur in the Mesozoic alongside brontosaurus, as the creationists claim; if it did, we would indeed have to question our evolutionary assumptions.

Scientists do ask questions about the pattern of evolutionary change. In particular, does the fossil record bear witness to the slow, continuous, gradual change envisioned by Darwin and supported by neo-Darwinists? Although still a matter of considerable debate, some form of punctuationalism is gaining increasing support among evolutionists. Scientists also ask questions about the process or mechanism of evolutionary change: for example, given a pattern of punctuational change, is Darwinian natural selection the best explanation for macroevolutionary trends?

The current debate is complicated because the concept of natural selection embraced by Darwinians has changed with the introduction of population genetics. Steven Stanley's concept of species selection (the differential survival of species) part of natural selection as formulated by Darwin and some modern biologists, but not as formulated by population geneticists focusing on selection operating within populations. Therefore, when Eldredge, Gould, and Stanley proclaim natural selection to be an inadequate explanation of macroevolutionary change, it is important to realize that they are talking about natural selection as mathematized, reformulated, and restricted to populational variation by population geneticists in the 1930s.

When a creationist such as Parker describes the putative failure of natural selection, however, it is to an audience that simplistically equates natural selection with evolution—an audience that does not know the difference between natural selection and species selection. Most students of scientific creationism know little about the debate between the phyletic gradualists and punctuationalists or that between proponents of Darwinian (nonrandom) and non-Darwinian (random) processes of change. And they will not learn what the debates are about from Parker and his colleagues.

"It's so utterly infuriating to find oneself quoted, consciously incorrectly, by creationists," Gould has said. "None of this controversy within evolutionary theory should give any comfort, not the slightest iota, to any creationist." Yet the scientific creationists, by misrepresenting the ongoing work of evolutionists, have helped the antievolutionary cause to gain more momentum than ever before in the twentieth century. Scientific creationists are widely viewed as learned scholars with impressive credentials, and more and more people are being persuaded that staggering evidence is on their side. Many scientists are baffled that such poor science can be so easily swallowed, and that creation is being taught as science in some schools around the country. Scientific creationism may be poor science, but it is powerful politics. And politically, it may succeed.

27

Science and Myth

JOHN MAYNARD SMITH

Recently, after giving a radio talk on Charles Darwin, I received through the post a pamphlet by Don Smith entitled "Why Are There Gays at All? Why Hasn't Evolution Eliminated Gayness Millions of Years Ago?" The pamphlet points to a genuine concern: the prevalence of homosexual behavior in our species is not understood and is certainly not something that would be predicted from Darwinian theory. Smith wrote the pamphlet because he believes the persecution of gays has been strengthened and justified by the existence of a theory of evolution that asserts that gays are unfit because they do not reproduce. He also believes that gays can be protected from future persecution only if it can be shown that they have played an essential and creative role in evolution. His argument is that, in evolution, novelty arises when individuals adopt mating habits different from those typical of their species: it is summed up on a button that reads "Sexual deviation is the mainspring of evolution."

I do not find this argument particularly persuasive, but that is not the point I want to make. I think Smith would have been better advised to have written, "If people despise gays because gayness does not contribute to biological fitness, they are wrong to do so. It would be as sensible to persecute mathematicians because an ability to solve differential equations does not contribute to fitness. A scientific theory—Darwinism or any other—has nothing to say about the value of a human being."

The point I am making is that Smith is demanding of evolutionary biology that it be a myth; that is, a story with a moral message. He is not alone in this. Elaine Morgan's book *The Descent of Woman* is an account of the origin of *Homo sapiens* that is intended to give mythical support to the women's movement by emphasizing the role of the female sex and, in particular, the mother–child bond. She claims, with reason, that many other accounts of human evolution have, perhaps unconsciously, placed undue emphasis on the role of males. Earlier, George Bernard Shaw wrote *Back to Methuselah* (1922) avowedly as an evolutionary myth, because he found in Darwinism a justification of selfishness and brutality and because he wished instead to support the Lamarckian theory of the inheritance of acquired characters, which he saw as justifying free will and individual endeavor.

We should not be surprised by Don Smith, Elaine Morgan, and Bernard Shaw. In all societies, people have constructed myths about the origins of the universe and of humans. The function of these myths is to define our place in nature and to give us a sense of purpose and value. Since Darwinism is, among other things, an account of human origins, is it any wonder that it is expected to carry a moral message?

The people and the objects that figure in a myth stand not only for themselves but also as symbols of other things. To some extent, myths and their symbolic components develop simply because human beings find it difficult to accept any input as meaningless. Shown an inkblot, we see witches, bats, and dragons. This refusal to accept input as mere noise lies at the root of divination by tarot cards, tea leaves, the livers or shoulders of animals, or the sticks of the *I Ching*. It may also account for the strangely late development of a mathematical theory of probability or of any scientific theory with a chance element. As anthropologist Dan Sperber has written, "Symbolic thought is capable, precisely, of transforming noise into formation."

Another—and in the present context, more important—function of myths is to give moral and evaluative guidance. Some mythmaking is quite conscious. In *Back to Methuselah,* for example, Shaw deliberately invented a story that would have the moral effect he desired. More usually, however, I suspect that a mythmaker conceives a story that moves him or her in a particular way—at its lowest, it reinforces prejudices, and at its highest, to borrow Aristotle's words, it evokes feelings of pity and fear. People repeat myths because they hope to persuade others to behave in certain ways.

This raises the question of why we use myths rather than simple statements of instruction. Why do we talk of King Alfred and the cakes, for example, instead of saying that people in important positions should be modest? Perhaps a story whose meaning has to be puzzled out or guessed at carries more conviction than a mere instruction. What we imagine is more important than what we are told.

Sometimes, I find it hard to discover how far people distinguish stories intended to give moral guidance from those meant simply to supply technical help. Confusion seems particularly likely to crop up when rituals are involved. For example, if, before going into battle, a man sharpens his spear and undergoes ritual purification (or, for that matter, cleans his rifle and goes to mass), he may regard the two procedures as equally efficacious. Indeed, they may well be so, one in preparing the spear and the other himself. If we regard the former as more practical, we do so only because we understand metallurgy better than psychology.

Despite the difficulty, most people do try to distinguish procedures and technical instructions that alter the external world from procedures and stories intended to alter our own state of consciousness or persuade us that certain things are right. Indeed, we take some trouble when educating our children to give hints about which category of information is being transmitted. For example, a surprisingly large proportion of the stories read aloud to children, particularly those with a moral message, are about talking animals or even talking steam engines. It is as if we wanted to be sure that the stories are not taken literally.

While such efforts may be successful in many spheres of human endeavor, the examples of Don Smith and Bernard Shaw show how hard it is for many people to separate science, and especially evolution, from myth. One reaction to this difficulty is to assert that there is no difference, that evolution theory has no more claim to objective truth than Genesis. Many scientists would be enraged by such an assertion, but rage is no substitute for argument. In the last century, it was widely held that the scientific method, conceived of as establishing theories by induction from observation, led to certain knowledge. Darwin and Einstein have robbed us of that certainty—or have liberated us from that prison. If, as Darwin showed, there is not a fixed and finite number of things in the universe, each with a knowable essence, then induction is logically impossible. Einstein, in turn, showed that what scientists had been most confident of—classical mechanics—was at worst false and at best a special case of a more general theory. After that twin blow, certain knowledge is something we can expect only at our funerals.

But it is one thing to admit that scientific knowledge cannot be certain and another to claim that there is no difference between science and myth. Karl Popper, perhaps the most influential contemporary philosopher of science, has told us that it was the impact of Einstein, and in particular the wish to distinguish Einstein's theory from those of Freud, Adler, and Marx, that led him to propose falsifiability as the criterion for separating science from pseudoscience. If a theory is scientific, he suggested, observations can be conceived of which, if they were accepted, would show the theory to be false. In contrast, he suggests that no conceivable pattern of human behavior could falsify Freudian theory.

Popper's views have been attacked, primarily on the grounds that there are no such things as theory-free observations. Every observation is subject to inter-pretation, conscious and unconscious. Consequently, there can never be certain grounds for rejecting a scientific theory, and hence the distinction between science and pseudoscience disappears.

This criticism seems to me largely to miss the point. If Popper were claiming that scientific knowledge were certain, then the impossibility of certain falsification would indeed be damaging. But he makes no such claim. He insists on two things. First, a scientific theory must assert that certain kinds of events cannot happen, so that the theory is falsified if these events are subsequently observed, and second, there is inevitably a logical asymmetry in any attempt to test a theory, so that a theory can be falsified but cannot be proved true by the acceptance of observation.

There is, however, a tide of ideas that would deny the distinction. The emotional force behind this tide derives, in part, from an entirely proper disgust at some of the consequences of technology in the modern world and, in part, from an equally proper wish to treat the ideas of other peoples as of equal value to our own. What is common to these two reactions is the conviction, which I share, that scientific theories are not the only kind of idea that we need. A frequently drawn corollary of this conviction, which I do not share, is that scientific ideas are not distinguishable from other ideas.

One source of the belief that science and myth can be lumped together lay, surprisingly, in Marxism, a philosophy that has led to two very different interpre-tations of science. One interpretation was pioneered by the British physicist J. D. Bernal. For Bernal, the crucial thing about science was that it made socialism possible by providing the techniques needed to satisfy people's wants. He saw science as being distorted under capitalism—for example, by being pressed into the service of military research; nevertheless, he does not appear to have thought that capitalism would prevent science from making progress toward an understand-ing of nature. In this, his views coincided with those of Marx himself, who largely excluded science from the set of ideas—for example, about religion, philosophy, and law—that he saw as reflecting the class interests of those who held them.

Thus Bernal regarded science as the greatest hope for the future and would have rejected any suggestion that science is indistinguishable from myth. However, another thread within Marxism has led to a different end. In 1931, the Russian B. Hessen argued that not only had Newton been influenced by the technical problems of his day (for example, gunnery and navigation) but also that the form his theory took reflected contemporary society. Such a view is perhaps more easily understood and more obviously true when applied to Darwin, whose theory did recognize in the natural world the processes of competition predominant in the society of his day. Indeed, both Darwin and Wallace stated that they borrowed their essential concept from the economist Malthus. If, then, this second thread of Marxist thinking argues, major scientific theories merely project onto nature features of contemporary society, they have more in common with myths than most scientists would readily accept.

Here, it seems to me, a crucial distinction must be made between the psychological sources of a theory and the testing of it. If Darwin's ideas, or Newton's, were accepted because they were socially appealing, then indeed science and myth would be indistinguishable. But I do not think that they were. They were accepted because of their explanatory power and ability to withstand experimental test. Of course, new ideas in science sometimes come from analogies with society, just as, in one scientific discipline, such as biology, they arise by analogy with others, such as physics and engineering. But what matters for the progress of science is not where the ideas come from but how they are treated.

Society influences the development of science through both the problems that seem worth solving and the resources available for their solution. I have little doubt that society also influences scientists, both as individuals and groups, by making some ideas seem worth pursuing, and others implausible or unpromising. For example, my own caution about applying to humans ideas drawn from a study of animal societies—a caution that contrasts with the enthusiasm of such scientists as E. O. Wilson and Richard Alexander—probably arose because I grew up under the shadow of Hitler and the Nazi theories of racial superiority and biological determinism and not because of anything internal to biology or sociology.

There is, however, a caricature inherent in the externalist view of science that I reject emphatically. This is the idea that we can evaluate a scientific theory by reference to the society in which it was born, or to the moral and political conclusions that might be drawn from it. Once accept that view and science is dead, as genetics died in Russia in 1948, when Stalin supported Lysenko's Lamarckian views against the Mendelians. Stalin took his position partly in the hope of quick returns in agricultural productivity and partly because Lysenko's belief in the inheritance of acquired characters seemed to accord better with Marxism than did the orthodox—and, as it happens, more nearly correct—Mendelian doctrine that hereditary characteristics are transmitted from parent to offspring by genes and that the genetic message is independent from changes induced in the body of the parent during its lifetime.

Today, the belief that there are no objective criteria whereby one can choose between rival theories (and hence, by implication, that one can allow one's prejudices full rein) derives largely, I think, from the work of Thomas Kuhn, although the conclusion is far from the one that he himself would wish to draw. Kuhn sees science as divided into "normal" and "revolutionary" periods. In a period of normal science, members of a scientific community agree about what assumptions can be made, what problems are worth solving, what will count as a solution, and what experimental methods should be used. Most important, they share a "paradigm," or set of exemplary solutions to problems, that can be used as a standard. Revolutions occur when, usually as a result of long-continued failure to solve certain problems within the accepted frame, a fundamentally new set of assumptions and procedures replaces the old.

All this seems to bear some resemblance to reality; it also bears some resemblance to Popper's remark that "there is much less accumulation of knowledge in science than there is revolutionary changing of scientific theories." The main difficulty lies in Kuhn's account of one "paradigm debate" in which the proponents "fail to make complete contact with each other's viewpoints" and in which they "see the world differently." Again, there is much in what he says. During my lifetime in science, I have engaged in too many arguments, in which I and my opponent have talked right past one another, not to recognize this.

The fallacy is to suppose that because two scientists are unable to understand each other fully, there is no rational way, given time, of settling the issue between them. With the passage of time, choosing between two theories or two methods of approach becomes easier. Eventually, one or the other approach is more successful in overcoming its difficulties or, as in the case of the particle and wave interpretations of light, a third theory is developed and subsumes them both. The trouble is that scientists must often commit themselves before the evidence is in. In Darwin's words, one must have "a theory by which to work." This is what gives an air of irrationality to the procedure and has led some people to suppose that the choice between scientific theories is arbitrary.

Consider an example. After the rediscovery of Mendel's laws in 1900, a debate broke out between the Mendelians and the earlier school of biometricians, headed by Karl Pearson. Pearson refused to accept the new theory, at least in part because it was incompatible with his previously held philosophical view that the business of science is merely to describe the world and not to imagine hypothetical entities such as genes. As the historian Bernard Norton has recently pointed out, Pearson understood that Mendelian inheritance could account for the phenomena of continuous variability, which he had been studying, but still rejected it on philosophical grounds—an interesting illustration of how philosophical preconceptions can be a poor guide to scientific practice. Yet, despite all this, no one today would doubt the utility either of Pearson's statistical methods or of Mendelian genetics.

If theories were genuinely incommensurable and rational choice between them impossible, progress in science would not be expected. Kuhn himself accepts the reality of scientific progress but only in the sense that the explanatory power of scientific theories has increased; he doubts whether it is sensible to say that science draws closer to the truth about what is "really there." I will return to this point in a moment.

Before leaving Kuhn, I want to suggest that, despite his insights, his insistence on a distinction between normal and revolutionary science, and on the incommensurability of paradigms, has been exaggerated. The major scientific revolution during my working life has been the rise of molecular biology, which has all the characteristics of a new "disciplinary matrix" in Kuhn's sense—new scientists, new problems, new experimental methods, new journals, new textbooks, and new culture heroes. But where was the incommensurability? I myself was raised in the older discipline of classical genetics and have never mastered the experimental methods of the new. Yet my almost immediate reaction to the Watson-Crick paper was that a mystery within my field had been cleared up. Those of us trained in classical genetics sometimes had difficulty in learning the new techniques, but there were few conceptual difficulties and no paradigm debate.

Perhaps the birth of molecular genetics was not a Kuhnian revolution. As it happens, I suspect that before we make much progress in developmental biology, a bigger conceptual revolution may be needed than the transition from classical to molecular genetics. All the same, if molecular biology could be born without the full panoply of a paradigm debate, where does that leave the concepts of normal and revolutionary science?

The history of genetics also forces us to look again at Kuhn's suggestion that progress in science is progress in explaining, but not progress in knowing what the world is really like. To me, the change from the concept of the gene as a Mendelian factor to the gene as a piece of a chromosome, and thence to the gene as a molecule of DNA, does look like progress in knowing what the world is like. But perhaps that is a question I should leave to the philosophers.

I would not have spent so much time discussing the difference between scientific theories and myths if the difference between them were obvious. Indeed, they have much in common. Both are constructs of the human mind, and both are intended to have a significance wider than the direct assertions they contain. Popper suggested falsifiability as the criterion distinguishing them, and I think he was right. However, we can often also distinguish them by their function: the function of a scientific theory is to account for experience—often, it is true, the rather esoteric experience emerging from deliberate experiment; the function of a myth is to provide a source and justification for values. What, then, should be the relation between them?

Three views are tenable. The first, sometimes expressed as a demand for "normative science," is that the same mental constructs should serve both as myths and as scientific theories. If I am right, this widely held view underlies the criticism of Darwinism from gays, from the women's movement, from socialists, and so on. It explains the preference expressed by some churchmen for "big bang" as opposed to "steady state" theories of cosmology. Although well intentioned, it seems to me pernicious in its effects. Applied to evolution theory, it means either that we must embrace Darwinism and draw from it the conclusion that gays are unnatural and social services wicked or that we must embrace Lamarckism whether or not the genetic evidence supports it. Normative science will be bad morality or bad science, and most probably both.

The second view is that we should do without myths and confine ourselves to science. This is the view I held at the age of twenty, but it really won't do. If, as I now believe, scientific theories say nothing about what is right but only about what is possible, we need some other source of values, and that source has to be myth, in the broadest sense of the term.

The third view, and I think the only sensible one, is that we need both myths and scientific theories, but that we must be as clear as we can about which is which. In essence, this was the view urged by the French molecular biologist Jaques Monod in *Chance and Necessity* (1972). Oddly, Monod was almost universally derided by his critics for arguing that one can derive values from science, when in fact he argued the precise opposite. His case was that there is no place in science for teleological, or value-laden, hypotheses. Yet, to do science, one must first be committed to some values—not least, to the value of seeking the truth. Since this value cannot be derived from science, it must be seen as a prior moral commitment, needed before science is possible. So far from values being derived from science, Monod saw science as depending on values.

Although I disagree with some aspects of his book, I agree with Monod on two basic points. First, values do not derive from science but are necessary for the practice of science. Second, we should distinguish as clearly as we can between science and myth. We should make this distinction, not because we could then discard the myths and retain only science, but because the roles they play are different. Scientific theories tell us what is possible; myths tell us what is desirable. Both are needed to guide proper action.

Index